DESIGN D&T MAKE IT !

systems and control technology

revised edition

Andy Biggs ■ Mike Hoffman ■ Tristram Shepard

First published in 1998 by:
Stanley Thornes (Publishers) Ltd

Second edition published in 2002 by:
Nelson Thornes Ltd
Delta Place
27 Bath Road
CHELTENHAM
GL53 7TH
United Kingdom

02 03 04 05 06 / 10 9 8 7 6 5 4 3 2 1

A catalogue record for this book is available from the British Library

ISBN 0 7487 6080 6

Designed by Carla Turchini
Illustrations by Tristram Ariss, Andrew Loft, Jane Cope and Hardlines

Printed and bound in Italy by Canale

Thanks are also due to Richard Calvert, Dave Mawson, Jet Mayor, Russ Jones and Roy Ballam for their contributions to the revised edition.

The publishers are grateful to the following for permission to reproduce photographs or other illustrative material:

AKA: p. 135 (bottom)
Bay Gen: p. 23 (bottom left)
Bentley: p. 14 (top)
Bridgeman Art Library: p. 23 (top right – Shaker table and sister's sewing chair, cherry wood, from New Lebanon, New York, 1815-20, American Museum, Bath, Avon)
British Standards Institute: p. 114 (top left)
Britstock – IFA: p. 132 (bottom – Bernd Ducke)
BTTG: pp. 102, 107 (right)
Channel 4 Learning: p. 98
Colorsport: p. 50 (middle left – Patrick Curtet)
Corel (NT): p. 23 (top left)
Cycling Plus: p. 108 (top)
Denford: p. 14 (bottom)
Dyson: p. 12
Economatics (Education) Ltd: pp. 35, 57, 58 (middle), 65 (bottom), 66 (top), 140
Ericsson: p. 45 (bottom left and right)
Eye Ubiquitous: p. 19 (Kevin Wilton)
Getty Images: pp. 30 (top – Hulton Deutsch), 32 (bottom – Hulton Deutsch), 36 (bottom – Ed Pritchard), 100 (middle – Phil Degginger), 106 (bottom – David Madison) (top – Hulton), 134 (top – Kevin Horan) (middle right)
Hemera Technologies Inc: pp. 8, 15, 17, 18 (top and bottom), 20, 25 (left), 88 (top), 146
Highways Agency: p. 95
Holt Studios International: p. 89 (bottom)
ICON Fitness Lifestyle: p. 88
IMAGE: p. 62
Image Bank: p. 56 (bottom left)
JCB: p. 103
John Walmsley: pp. 23 (top), 50 (top right)
Julian Molloy: p. 131 (top)
JVC: p. 27 (top)
Last Resort Picture Library: pp. 16, 40 (top), 59, 62 (bottom middle right)
LEGO Dacta UK: pp. 13, 60 (middle right), 61, 136
LEICA: p. 119 (bottom)
Leslie Garland: pp. 50 (bottom), 52 (bottom), 131 (bottom)
London Aerial Photolibrary: p. 105 (top)
Martyn Chillmaid: pp. 25 (top), 27 (middle), 33, 34, 37, 38 (top), 40 (bottom), 42 (middle), 44, 48 (right), 49, 51 (middle), 52, 56 (top left), 57, 58 (top and middle), 62 (bottom right and left), 76 (top left and bottom), 77, 78, 79, 81, 98, 99, 102, 106 (top), 107 (left), 113 (left), 114 (bottom), 115 (middle and bottom), 120 (top), 122 (bottom), 125 (middle), 126 (middle and right), 127, 128, 130, 131, 132, 135 (top), 139, 141 (left and right), 148
MIRA: pp. 104 (top left), 105 (bottom right)
Moulinex Swan: p. 30 (bottom)
NASA: p. 22 (top)
National Exhibition Centre: p. 41
PA News Photolibrary: p. 25 (bottom)
Philip Poole: pp. 59 (top), 134 (bottom)
Photodisc (NT): p. 27 (bottom)
Rapid Electronics: pp. 45 (top and bottom), 47 (top), 74, 77
Rex Features: pp. 23 (bottom right – Adrian Brooks), 147 (bottom left and right)
Robert Harding Picture Library: p. 26 (bottom)
Rolls Royce plc: p. 135 (top right)
Rover Group: pp. 104 (top right), 105 (middle, middle right and bottom left)
Science Photolibrary: pp. 26 (middle right – James Holmes/Rover), 64 (top – David Parker, bottom – Rosenfeld Images Ltd), 138 (top – Bruce Iverson) Science and Society Picture Library: p. 36 (top)
Simon Phillips: pp. 45 (top and middle), 56 (bottom right), 106 (bottom)
Sony: p. 9
Stockbyte (NT): p. 147 (top)
Stockmarket: p. 135 (top left)
Studio Collins: p. 89 (top left)
Telegraph Colour Library: p. 150 (M Simpson)
Topham Picturepont: p. 18 (middle – UPP)
Toyota: p. 24
Travel Ink: pp. 84 (top – David Toase), 85 (David Toase)
UKAEA Technology: p. 111 (middle left)
Yale: pp. 50, 88

Contents

Introduction

Welcome to Design & Make It: Systems and Control. *This book has been written to support you as you work through your GCSE course in Design and Technology. It will help guide you through the important stages of your coursework, and assist your preparation for the final examination paper.*

Beyond GCSE

There are good opportunities for skilled people to work in industry. Such people need to be flexible, good communicators, willing to work in teams, and to be computer literate. Further courses and training opportunities are available at various levels which you might like to find out more about.

Making It!

Whatever your project, remember that the final realisation is particularly important. It is not enough just to hand in your design folder. You must have separate products which you have made. The quality of your final realisation must be as high as possible, as it counts for a high proportion of the marks.

During your course you will need to develop technical skills in using electronic and/or pneumatic components and your constructional skills. This is something you can't do just by reading a book! The best way is to watch carefully as different techniques and procedures are demonstrated to you, and practise them as often as possible.

The Written Paper

Four pages of examination-style questions are included to help give you confidence in tackling the types of questions that you will be set.

How to Use this Book

There are two main ways you might use the book.

1   After the introductions to the two themes you could undertake one or more of the projects which follow, depending on the 'focus' option you are doing. This will ensure coverage of the systems 'core'. You will need to consult other specialist texts to ensure you cover the syllabus of your focus area fully.

2   Undertake alternative projects to one or more of those provided. Refer to those pages which cover the project guide, and the specific areas of knowledge and understanding defined in the examination syllabus.

Contents

Project guide
The Project Guide summarises the main design skills you will need for extended project work. Refer back to these pages throughout the course.

The book then provides a general introduction to the concepts of systems and sub-systems. Five case-studies provide reference material for subsequent project work.

The projects
Two themes are provided to introduce five projects which each contain knowledge and understanding pages (e.g. Pneumatics, Sensors, Mechanisms, etc), which cover the Systems and Control Core.

'Focus' project suggestions
Finally two alternative outline project suggestions are provided as a guideline for extended 'Focus' project work. Refer back to the earlier Project Guide to help ensure you are covering the necessary developmental stages and recording their progress in the way examiners will be looking for.

■ ACTIVITY

Make sure that as part of your project folio you include evidence of having completed a number of short-term focused tasks, as suggested in the Activities.

IN YOUR PROJECT

The 'In Your Project' lists will help you to think about how you could apply the content of the page to your current work.

KEY POINTS

Use the 'Key Points' lists to revise from when preparing for the final examination paper.

Choosing and Starting Projects

Identifying suitable design and make projects for yourself is not easy. A carefully chosen project is much more likely to be interesting and easier to complete successfully. Investing time and effort in choosing a good project makes progress a lot easier later on.

Project Feasibility Studies

Make a start by making a list of:

▷ potential local situations/environments you could visit where you could do some research into the sort of things people there might need (e.g. a local playgroup, a small business, a hospital or sports centre, etc.)

▷ people you know outside school who might be able to help by providing information, access and/or advice.

The next stage is to get up and get going. Arrange to visit some of the situations you've listed. Choose the ones which you would be interested in finding out some more about. Make contact with the people you know, and get them interested in helping you. Tell them about your D&T course, and your project.

For each possible situation you should:

▷ visit the situation or environment
▷ make initial contact with those whose help you will need.

With a bit of luck, after you've done the above you should have a number of ideas for possible projects.

Try to identify what the possible outcomes of your projects might be – not what the final design would be, but the form your final realisation might take, e.g. a working production model, a series of system diagrams, plans and elevations, etc. Think carefully about the following:

▷ Might it be expensive or difficult to make?
▷ Do you have access to the tools and materials which would be required?
▷ Will you be able to find out how it could be manufactured?
▷ Does the success of the project depend on important information you might not be able to get in the time available?
▷ Are there good opportunities for you to use ICT?

the home

energy

the natural environment

the high street

transport

communications

clothing

leisure

security

food

health

education

starting points

There are a number of different ways in which you might start a project. Your teacher may have:

● told you exactly what you are required to design
● given you a range of possible design tasks for you to choose from
● left it up to you to suggest a possible project.

If you have been given a specific task to complete you can go straight on to page 8.

If you are about to follow one of the main units in this book, you should go straight to the first page of the task.

If you have been given a number of possible tasks to choose from you should go straight to the section on page 7 entitled 'Making your Choice'.

However, if you need to begin by making some decisions about which will be best task for you, then the first stage is to undertake some project feasibility studies as described on this page.

Making Your Choice

For each of your possible projects work through pages 8-9 (Project Investigation) and try planning out a programme of research.

Look back over your starting questions and sources of information:

▷ Could you only think up one or two areas for further research?
▷ Did you find it difficult identifying a range of sources of information?

If this has been the case, then maybe it is not going to prove to be a very worthwhile project.

Ideally, what you're looking for is a project which:

▷ is for a nearby situation you can easily use for research and testing
▷ you can get some good expert advice about.
▷ shows a good use of ICT.

It is also important that your expected outcome:

▷ will make it possible for you to make and test a prototype
▷ will not be too difficult to finally realise.

Finally, one of the most important things is that you feel interested and enthusiastic about the project!

don't forget...

A very important consideration is the testing of a prototype of some sort, and of your final design. How would you be able to do this? Could ICT be used?

Remember it's important that what you design is suitable for production, even if only in small numbers. It can't be just a one-off item. You will need to show some plans for your product to be factory made.

Don't forget to record all your thoughts and ideas about these initial stages of choosing and starting your project.

In your project folio provide a full record of the ideas you reject, and the reasons why. This helps provide important evidence of your decision-making skills, and of the originality of your project. Communication skills are important.

If you come up with more than one good idea, find out how many projects you have to submit at the end of your course. You might be able to do one or more of your other ideas at a later date.

Make sure you discuss your project ideas with a teacher.

in my design folder

✓ My project is to design a...
✓ I am particularly interested in...
✓ I have made a very good contact with...
✓ My prototype can be tested by...
✓ My final outcome will include...
✓ I could use ICT to...

Project Investigation

You will need to find out as much as you can about the people and the situation you are designing for. To do this you will need to identify a number of different sources of information to use for your research. Research must be relevant, and is likely to continue throughout the task.

Starting Questions

Make a list of questions you will have to find answers to.

You should find the following prompts useful:

Why...?

When...?

Where...?

What...?

How many...?

How often...?

How much...?

Sources of Information

Next, carefully consider and write down the potential sources of information you might be able to use in order to discover the answers to your starting questions.

Work through the research methods on the next page. Be sure to give specific answers as far as possible (i.e. name names).

Across your research you will need to aim to obtain overall a mixture of:

▷ factual information: e.g. size, shape, weight, cost, speed, etc.
▷ information which will be a matter of opinion: i.e. what people think and feel about things, their likes and dislikes, what they find important, pleasing, frustrating, etc.

don't forget...

> Write down what you need to find out more about, and how you could obtain the information.

> Make sure your research work is clearly and attractively presented.

> Identify a number of sources of information, and undertake the research, remembering to record what you discover.

> You need to identify a number of sources of information (e.g. user research, existing solutions, expert opinion, information search). The wider range of methods you use, the more marks you will get.

in my design folder

✓ The key things I need to find out about are...
✓ The research methods I am going to use are...
✓ I will be talking to the following people about my project...
✓ I will need to have it all completed by...
✓ I will use ICT to...

Research Methods

User Research

Which people could you observe and consult who are directly involved in the situation? To what extent do you consider that you will be able to find out about:

- the things they do
- the way in which they do them.

As well as asking individuals, you could also undertake a small survey or questionnaire.

User Trips

How can you record your own impressions of the situation in which the product will be used? Are there any activities you could try out for yourself to gain first-hand experience? Have you had any previous similar experiences?

Similar Situations

Do you know of any other comparative circumstances in which people are in similar situations, and which might help provide insight and ideas?

Expert Opinion

Are there any people you know of who could give you expert professional advice on any aspects of the situation? If you don't know of anyone, how might you set about finding somebody?

Information Search

Has any information about the situation, or a similar situation, been documented already in books, magazines, TV programmes, the Internet, or CD-ROM? If you don't already know that such information exists, where could you go to look for it? Don't forget to consider the possibility of using information stored on a computer database.

Existing Products

Make a study of solutions to similar problems which already exist. Describe how they work and discuss how well they work.

In Conclusion

When most of your investigation work has been completed you will need to draw a series of conclusions from what you have discovered. What have you learnt about the following things:

▷ What sort of people are likely to be using the product?
▷ Where and when will they be using it?
▷ What particular features will it need to have?
▷ How many should be made?

Of all the research methods, user-research tends to be the most effective and useful, so you are highly recommended to include some in your investigation. Some form of personal contact is essential to a really successful project.

It is also highly advisable to conduct some form of questionnaire. If you have not done one to submit as part of your coursework, make sure that you will have the opportunity to do so this time.

It isn't necessary to use all the research methods in any one project, but you certainly must show that you have tried a selection of them.

in my design folder

✓ From my research I found out...
✓ I have discovered that...
✓ My conclusions are...
✓ I have kept my research relevant by...
✓ I found ICT helpful when...

From Design Specification to Product Specification

A design specification is a series of statements that describe the possibilities and restrictions of the product. A product specification includes details about the features and appearance of the final design, together with its materials, components and manufacturing processes.

Use a word processor to draft and finalise your design specification.

Writing a Design Specification

The **design specification** is a very important document. It describes the things about the design which are fixed and also defines the things which you are free to change.

The conclusions from your research should form the basis of your design specification. For example, if in your conclusions you wrote:

'From the measurements I made of the distance at which a number of people could see a seven-segment display, I discovered most could read a standard display from 3 m.'

In the specification you would simply write:

'The display should be able to be read from 3 m.'

The contents of the specification will vary according to the particular product you are designing, but on the next page is a checklist of aspects to consider. Don't be surprised if the specification is quite lengthy. It could easily contain 20 or more statements.

Fixing It

Some statements in the specification will be very specific, e.g.: *'The toy must be red.'*

Other statements may be very open ended, e.g.: *'The toy can be any shape or size.'*

Most will come somewhere in between, e.g.: *'The toy should be based on a vehicle of some sort and be mechanically or electronically powered.'*

In this way the statements make it clear what is already fixed (e.g. the colour), and what development is required through experimentation, testing, and modification (e.g. shape, size, vehicle-type and method of propulsion).

Writing a Product Specification

After you have fully developed your product you will need to write a final more detailed **product specification**. This time the precise statements about the materials, components and manufacturing processes will help ensure that the manufacturer is able to make a repeatable, consistent product.

Your final product will need to be evaluated against your design specification to see how closely you have been able to meet its requirements, and against your product specification to see if you have made it correctly.

don't forget...

You might find it helpful to start to rough out the design specification first, and then tackle the conclusions to your research. Working backwards, a sentence in your conclusion might need to read:

'From my survey, I discovered that young children are particularly attracted by bright primary colours.'

It's a good idea to use a word processor to write the specification. After you've written the design specification new information may come to light. If it will improve the final product, you can always change any of the statements.

Make sure you include as much numerical data as possible in your design specification. Try to provide data for anything which can be measured, such as size, weight, quantity, time and temperature.

Specification Checklist

The following checklist is for general guidance. Not all topics will apply to your project. You may need to explore some of these topics further during your product development.

Use and performance
Write down the main purpose of the product – what it is intended to do. Also define any other particular requirements, such as speed of operation, special features, accessories, etc. Ergonomic information is important here.

Size and weight
The minimum and maximum size and weight will be influenced by things such as the components needed and the situation the product will be used and kept in.

Generally the smaller and lighter something is the less material it will use, reducing the production costs. Smaller items can be more difficult to make, however, increasing the production costs.

Appearance
What shapes, colours and textures will be most suitable for the type of person who is likely to use the product? Remember that different people like different things.

These decisions will have an important influence on the materials and manufacturing processes, and are also crucial to ensure final sales.

Safety
A product needs to conform to all the relevant safety standards.
- Which of them will apply to your design?
- How might the product be mis-used in a potentially dangerous way?
- What warning instructions and labels need to be provided?

Conforming to the regulations can increase production costs significantly, but is an area that cannot be compromised.

Manufacturing cost
This is concerned with establishing the maximum total manufacturing cost which will allow the product to be sold at a price the consumer or client is likely to pay.

The specification needs to include details of:
- the total number of units likely to be made
- the rate of production and, if appropriate
- the size of batches.

Maintenance
Products which are virtually maintenance free are more expensive to produce.
- How frequently will different parts of the product need to be maintained?
- How easy does this need to be?

Life expectancy
The durability of the product has a great influence on the quantity of materials and components and the manufacturing process which will need to be used.

How long should the product remain in working order, providing it is used with reasonable care?

Environmental requirements
In your specification you will need to take into account how your product can be made in the most environmentally friendly way. You might decide to:
- specify maximum amounts of some materials
- avoid a particular material because it can't be easily recycled
- state the use of a specific manufacturing process because it consumes less energy.

Other areas
Other statements you might need to make might cover special requirements such as transportation and packaging.

in my design folder

✓ My design will need to...
✓ The requirements of the people who will use it are...
✓ It will also need to do the following...
✓ It will be no larger than...
✓ It will be no smaller than...
✓ Its maximum weight can be...

✓ It should not be lighter than...
✓ The shapes, colours and textures should...
✓ The design will need to conform to the following safety requirements...
✓ The number to be made is...
✓ The following parts of the product should be easily replaceable...
✓ To reduce wastage and pollution it will be necessary to ensure that...

Generating and Developing Ideas (1)

When you start designing you need lots of ideas – as many as possible, however crazy they might seem. Then you need to start to narrow things down a bit by working in more detail and evaluating what you are doing.

Wherever possible consider using a computer to experiment with your ideas, and to analyse and present your findings.

First Thoughts

Start by exploring possibilities at a very general level. Spend time doing some of the following:

▷ Brainstorming, using key words and phrases or questions which relate to the problem.
▷ Completing spider-diagrams which map out a series of ideas.
▷ Using random word or object-association to spark off new directions.
▷ Thinking up some good analogies to the situation (i.e. what is it like?).
▷ Work from an existing solution by changing some of the elements.
▷ Experimenting with some materials.

Continue doing this until you have at least two or three possible approaches to consider. Make sure they are all completely different, and not just a variation on one idea.

Go back to your design specification. Which of your approaches are closest to the statements you made? Make a decision about which idea to take further, and write down the reasons for your choice.

As you work through this section it is important to remember the following sequence when considering potential solutions:
● record a number of different possibilities
● consider and evaluate each idea
● select one approach as the best course of action, stating why.

There are lots of different drawing techniques which you can use to help you explore and develop your ideas, such as plans, elevations, sections, axonometrics, perspectives, etc.
Try to use as rich a mixture of them as possible. At this stage they should really be 'rough', rather than 'formal' (i.e. drawn with a ruler). Colour is most useful for highlighting interesting ideas.

don't forget...

As usual, it is essential to record all your ideas and thoughts.

Much of your work, particularly early on, will be in the form of notes. These need to be neat enough for the examiner to be able to read.

Drawings on their own do not reveal very well what you had in mind, or whether you thought it was a good idea or not. Words on their own suggest that you are not thinking visually enough. Aim to use both sketches and words.

Communicating your ideas

Communicating your ideas clearly and effectively through labelled drawings will help you to:

▷ visualise the ideas that you have in your head;

▷ record your ideas and your reasons for developing the product the way you have;

▷ explain your ideas to others, including your teacher and the examiner.

Sketching

The drawing technique you use needs to be quick and clear. Sketches should be freehand - rulers should not be used as they take time and can restrict your design work to straight-line shapes.

Draw in 3D or use plans and elevations. Use colour and shade only if it helps to explain your ideas, not just to decorate the drawing. Use written notes to help explain and comment on your ideas.

The LED lights up when the object is gripped

The switch will control a motor to close the grips

This section of the grabber will connect the switch to the grips

The batteries can be replaced here

System Diagrams

Use system and circuit diagrams to show the development of your design.

Electronic circuit diagrams should be drawn carefully using the correct symbols. PP7303 sets the standard for electrical and electronic graphical symbols.

Technical Drawings

Technical drawings will help you to:

▷ produce accurate PCB masks;

▷ position components sensibly on a schematic diagram;

▷ test that the parts fit together.

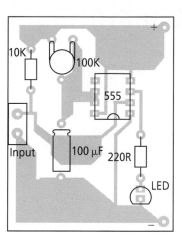

For complicated details it may be necessary to draw things twice full-size (2:1), or if the product is large, half full-size (1:2). PCB masks should always be full-size (1:1).

You will need to prepare accurate technical drawings of the casing of your product before you make the final version. They should be good enough for someone else to be able to make it from.

Plans and elevations drawn together are known as **orthographic projection**. Make sure you follow the conventions for dimensioning. Sections through the product can often help to explain constructional details, as can exploded drawings.

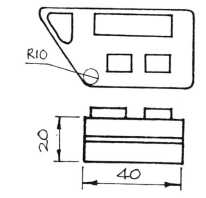

Circuit Design - Development

Generating and Developing Ideas (2)

CAD-CAM can be extremely useful at this stage. Work towards making at least one prototype to test out some specific features of your design. Record the results and continue to refine your ideas as much as you can. Sorting out the final details often requires lots of ideas too.

To find out more about 2D and 3D CAD programs, go to:
www.bentley.com
www.adobe.com

CAD / CAM

Computer Aided Design (CAD) and **Computer Aided Manufacture (CAM)** are terms used for a range of different ICT applications that are used to help in the process of designing and making products.

CAD is a computer-aided system for creating, modifying and communicating ideas for a product or components of a product.

CAM is a broad term used when several manufacturing processes are carried out at one time aided by a computer. These may include process control, planning, monitoring and controlling production.

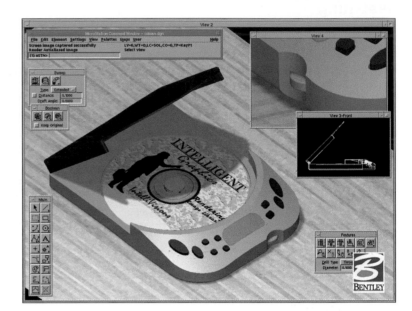

CAM or CNC systems may be used for rapid prototyping or modelling in 'soft' materials, the manufacture of small parts and moulds, and the production of templates, stencils or printouts on paper, card, vinyl, etc.

You may not have CAD-CAM or CNC systems in your workshops but you can still indicate how batch production of your product might make use of CNC machines like lathes, milling machines, flatbed routers, spindle moulders, injection and extrusion machines, etc.

CIM

Computer Integrated Manufacture is extensively used by industry. This is where computers are used to control the whole production process from materials input and handling, to pick and placement, to machining to construction and possible packaging.

ICT ➡

Wherever possible consider using a CAD program. Designing on-screen happens very quickly and little evidence of change is seen compared with drawing on paper. You need to develop ways of recording your thinking and collect evidence of any developments.

- Make sure you print out the various stages you work through, or keep a copy on disk

- Where CAM/CNC is used you must record all your programming and evidence of machine set-ups in your project folder.

As you develop your ideas, make sure you are considering the following:
- Design – aesthetics, ergonomics, marketing potential, etc.
- User requirements – functions and features.
- Technical viability – if it could be made.
- Manufacturing potential – how it could be made in quantity.
- Environmental concerns – if it can be reused, recycled, etc.?

Models are simplified versions of intended products. Use words, numbers, drawings and 3D representations of your ideas to help you develop and evaluate your designs as they progress.

At some stage you will need to move off the drawing board and try some things out in three dimensions using real materials, circuit components and kits.

Planning and Making Prototypes

At some stage you will be in a position to bring your ideas into sharper focus by making some form of mock-up or prototype. Think carefully about exactly what aspect of your idea you want to test out and about the sort of model which will be most appropriate.

Whatever the form of your final outcome, the prototype might need to be:

▷ two-dimensional
▷ three-dimensional
▷ made using a prototype board or system kit
▷ made using different materials at a different scale.

Try to devise some objective tests to carry out on your prototype involving measuring something. Don't rely just on people's opinions. Write up the circumstances in which the tests were undertaken, and record your results.

Write down some clear statements about:

▷ what you wanted the prototype to test
▷ the experimental conditions
▷ what you discovered
▷ what decisions you took about your design as a result.

Following your first prototype you may decide to modify it in some way and test it again, or maybe make a second, improved version from scratch. Make sure you keep all the prototypes you make, and ideally take photographs of them being tested perhaps using a digital camera.

Sometimes you will need to go back to review the decisions you made earlier, and on other occasions you may need to jump ahead for a while to explore new directions or to focus down on a particular detail. Make sure you have worked at both a general and a detailed level.

in my design folder

✓ I chose this idea because...
✓ I developed this aspect of my design by considering...
✓ To evaluate my ideas I decided to make a prototype which...
✓ The way I tested my prototype was to...
✓ What I discovered was...
✓ As a result I decided to change...
✓ I used ICT to...

15

Planning the Making and the Manufacturing

The final realisation is very important. It presents your proposed design solution rather than the process you used to develop it. Careful planning is essential. You will also need to be able to explain how your product could be manufactured in quantity.

How many?

What you have designed should be suitable for manufacture. You should discuss with your teacher how many items you should attempt to make. This is likely to depend on the complexity of your design and the materials and facilities available in your workshops. It may be that you only make one item, but also provide a clear account of how a quantity of them could be manufactured.

keeping a record

Write up a diary record of the progress you made while making. Try to include references to:

- things you did to ensure safety
- the appropriate use of materials
- minimising wastage
- choosing tools
- practising making first
- checking that what you are making is accurate enough to work
- asking experts (including teachers) for advice explaining why you had to change your original plan for making.

A Plan of Action

Before you start planning you will need to ensure that you have an orthographic drawing of your design (see page 13). This will need to include all dimensions and details of the materials to be used.

Ideally there should also be written and drawn instructions which would enable someone else to make up the design from your plans.

Next work out a production flow chart as follows.

1. List the order in which you will make the main parts of the product. Include as much detail as you can

2. Divide the list up into a number of main stages, e.g. gathering materials and components, preparing (i.e. marking out, cutting), assembling, finishing.

3. Identify series of operations which might be done in parallel.

4. Indicate the time scale involved on an hourly, daily and weekly basis.

Consider the use of templates and jigs to help speed things up. Other possibilities include the use of moulds or setting up a simple CAM system to produce identical components.

don't forget...

You may find you have to change your plans as you go. There is nothing wrong with doing this, but you should explain why you have had to adjust your schedule, and show that you have considered the likely effect of the later stages of production.

Quality Counts

As your making proceeds you will need to check frequently that your work is of acceptable quality. How accurately will you need to work? What tolerances will be acceptable (see page 144)? How can you judge the quality of the finish?

If you are making a number of identical items you should try and work out ways of checking the quality through a sampling process (see page 144).

Making

While you are in the process of making you must ensure that the tools and materials you are using are the correct ones. Pay particular attention to safety instructions and guidelines.

Try to ensure that you have a finished item at the end, even if it involves simplifying what you do.

Aim to produce something which is made and finished as accurately as possible. If necessary you may need to develop and practise certain skills beforehand.

Planning for Manufacture

Remember to use a wide range of graphic techniques to help plan and explain your making.

Don't forget that there is also a high proportion of marks for demonstrating skill and accuracy, overcoming difficulties and working safely during the making.

What needs to be done by:
• next month
• next week
• next lesson
• the end of this lesson?

Manufacturing matters

Try asking the following questions about the way your design might be made in quantity:
- What work operation is being carried out, and why? What alternatives might there be?
- Where is the operation done, and why? Where else might it be carried out?
- When is it done, and why? When else might it be undertaken?
- Who carries it out, and why? Who else might do it?
- How is it undertaken, and why, How else might it be done?

Remember that manufacturing is not just about making things. It is also about making them better by making them:
- simpler • quicker
- cheaper • more efficient
- less damaging to the environment.

Try to explain how your product would be manufactured in quantity. Work through the following stages:

1 Determine which type of production will be most suitable, depending on the number to be made.
2 Break up the production process into its major parts and identify the various sub-assemblies.
3 Consider where jigs, templates and moulding processes could be used. Where could 2D or 3D CAM be effectively used?
4 Make a list of the total number of components and volume of raw material needed for the production run.

5 Identify which parts will be made by the company and which will need to be bought in ready-made from outside suppliers.
6 Draw up a production schedule which details the manufacturing process to ensure that the materials and components will be available exactly where and when needed. How should the workforce and workspace be arranged?

7 Decide how quality control systems will be applied to produce the degree of accuracy required.
8 Determine health and safety issues and plan to minimise risks.
9 Calculate the manufacturing cost of the product.
10 Review the design of the product and manufacturing process to see if costs can be reduced.

in my design folder

✓ I planned the following sequence of making...
✓ I had to change my plan to account for...
✓ I used the following equipment and processes...
✓ I paid particular attention to safety by...
✓ I monitored the quality of my product by...
✓ My product would be manufactured in the following way...

Testing and Evaluation

You will need to find out how successful your final design solution is. How well does it match the design specification? How well have you worked? What would you do differently if you had another chance?

Alarm system

As you work through your project you will regularly carry out testing and evaluation. For example:

▷ analysing and evaluating the research material you collected
▷ evaluating and testing carried out or existing products
▷ evaluating initial sketch ideas or prototypes and models in order to make the right decisions about which to develop further
▷ assessing the quality of your making as you go along.

Last of all, you must test and evaluate your final solution.

Virtual pets

Testing the Final Solution

To find out how successful your design is you will need to test it out. Some of the ways in which you might do this are by:

▷ trying it out yourself
▷ asking other people to use it
▷ asking experts what they think about it.

As well as recording people's thoughts, observations and opinions, try to obtain some data: how many times it worked, over what periods of time, within what performance limits, etc?

To help you decide what to test, you should look back to the statements in your design specification and focus on the most important ones. If for example the specification stated that a three-year-old child must be able to operate it, try and find out if they can. If it must be a colour which would appeal to young children, devise a way of finding out what age ranges it does appeal to.

You need to provide evidence to show that you have tested your final design out in some way. Try to ensure that your findings relate directly to the statements in your original specification. Include as much information and detail as you can.

What methods could you use to test the success of the control systems used in these products?

Heart rate monitor

don't forget...

Don't be too surprised or worried if your design isn't perfect – the important thing is that you can identify what needs improving. Can you make some simple suggestions about how it might be improved?

Final Evaluation

There are two things you need to discuss in the final evaluation: the quality of the product you have designed, and the process you went through while designing it.

The product

How successful is your final design? Comment on things like:

▷ how it compares with your original intentions as stated in your design specification
▷ how well it solves the original problem
▷ the materials you used
▷ what it looks like
▷ how well it works
▷ what a potential user said
▷ what experts said
▷ whether it could be manufactured cheaply enough in quantity to make a profit
▷ the effective use of ICT to assist reproduction or manufacture
▷ the extent to which it meets the requirements of the client, manufacturer and the retailer
▷ the ways in which it could be improved.

Justify your evaluation by including references to what happened when you tested it.

The process

How well have you worked? Imagine you suddenly had more time, or were able to start again, and consider:

▷ Which aspects of your investigation, design development work and making would you try to improve, or approach in a different way?
▷ What did you leave to the last minute, or spend too much time on?
▷ Which parts are you most pleased with, and why?
▷ How well did you make the final realisation?
▷ How effective was your use of ICT? How did it enhance your work?

If you had more time:

● what aspects of the product would you try to improve? (refer to your evaluation if you can).
● how would you improve the way you had researched, developed, planned and evaluated your working process?

in my design folder

What do you think you have learnt through doing the project?
✓ Comparison of my final product specification with my design specification showed that...
✓ The people I showed my ideas (drawings and final product) to said...
✓ I was able to try my design out by...
✓ I discovered that...
✓ I could improve it by...
✓ I didn't do enough research into...
✓ I spent too long on...
✓ I should have spent more time on...
✓ The best aspect is...
✓ I have learnt a lot about...

Try to identify a mixture of good and bad points about your final proposal and method of working. You will gain credit for being able to demonstrate that you are aware of weaknesses in what you have designed and the way that you have designed it.

Don't forget to write about both the product and the process.

If people have been critical of aspects of your design, do you agree with them? Explain your response.

Remember that evaluation is on-going. It should also appear throughout your project whenever decisions are made. Explain the reasons behind your actions.

Project Presentation

The way you present your project work is extremely important. Remember you won't be there to explain it all when it's being assessed! You need to make it as easy as possible for an examiner to see and understand what you have done.

Telling the Story

All your investigation and development work needs to be handed in at the end, as well as what you have made. Your design folder needs to tell the story of the project. Each section should lead on from the next, and show clearly what happened next, and explain why. Section titles and individual page titles can help considerably.

There is no single way in which you must present your work, but the following suggestions are all highly recommended:

▷ Securely bind all the pages together in some way. Use staples or treasury tags. There is no need to buy an expensive folder.
▷ Add a cover with a title and an appropriate illustration.
▷ Make it clear which the main sections are.
▷ Add titles or running headings to each sheet to indicate what aspect of the design you were considering at that particular point in the project.

Remember to include evidence of ICT work and other Key Skills. Carefully check through your folder and correct any spelling and punctuation mistakes.

Presenting your Design Project Sheets

▷ Always work on standard-size paper, either A3 or A4.

▷ Aim to have a mixture of written text and visual illustration on each sheet.

▷ You might like to design a special border to use on each sheet.

▷ Include as many different types of illustration as possible.

▷ When using photographs, use a small amount of adhesive applied evenly all the way around the edge to secure them to your folder sheet.

▷ Think carefully about the lettering for titles, and don't just put them anywhere and anyhow. Try to choose a height and width of lettering which will be well balanced on the whole page. If the title is too big or boldly coloured it may dominate the sheet. If it is thin or light it might not be noticed.

don't forget...

Presentation is something you need to be thinking about throughout your project work.

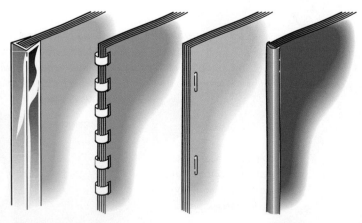

Binding methods

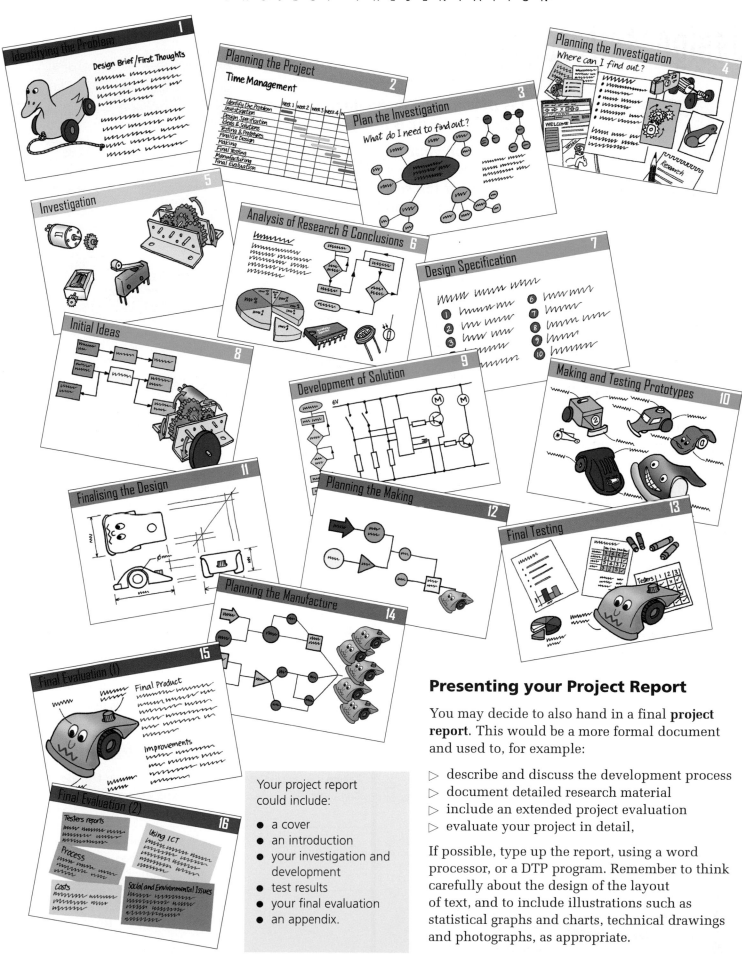

Presenting your Project Report

You may decide to also hand in a final **project report**. This would be a more formal document and used to, for example:

▷ describe and discuss the development process
▷ document detailed research material
▷ include an extended project evaluation
▷ evaluate your project in detail,

If possible, type up the report, using a word processor, or a DTP program. Remember to think carefully about the design of the layout of text, and to include illustrations such as statistical graphs and charts, technical drawings and photographs, as appropriate.

Your project report could include:

● a cover
● an introduction
● your investigation and development
● test results
● your final evaluation
● an appendix.

Design Matters (1)

What is Design and Technology, how has it changed, and why is it important?

As you develop your ideas for products you will often need to make important decisions about the social, moral, cultural and environmental impact of your product.

All Change, Please!

Technology is about making our lives easier and more comfortable. It involves using our knowledge and understanding of science and human needs together with our skills of making.

Advances in technology change what needs to be done and the way in which we need to do it.

The rate of change of technological achievement has rapidly increased in recent years. The Industrial Revolution and digital electronics are two things that have had a major impact on our lives.

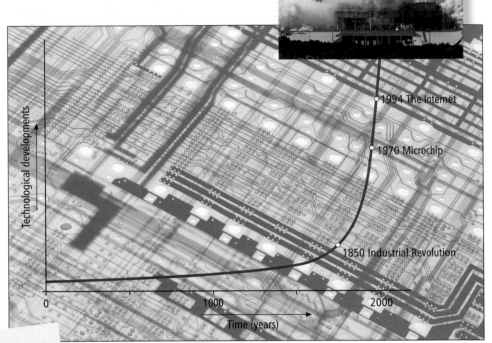

'...if the last 50,000 years of Man's existence were divided into lifetimes of approximately sixty-two years each, there have been about 800 such lifetimes. Of these 800, fully 650 were spent in caves.

Only during the last seventy lifetimes has it been possible to communicate effectively from one lifetime to another – as writing made it possible to do.

Only during the last six lifetimes did masses of people ever see a printed word. Only during the last four has it been possible to measure anything with precision.

Only in the last two has anyone anywhere used an electric motor. And the overwhelming majority of all the material goods we use in daily life today have been developed within the present, the 800th lifetime.'

From 'Future Shock', Alvin Toffler.

■ ACTIVITY

Read the passage on the left and study the graph above. Identify an activity and identify examples which show how technological development has changed our lives. Add your examples to a table like the one below. In some cases the impact may not necessarily be advantageous.

Activity	Then	Now	Impact
Telling the time	Sun dial	Digital watch	Everyone can tell the time with a high degree of accuracy.
Cooking	Open fire	Micro-wave	Food can be re-heated from frozen and heated very quickly.

Design and Technology is about improving people's lives by designing and making the things they need and want. But different people have different needs: what is beneficial to one person can cause a problem for someone else, or create undesirable damage to the environment.

A new design might enable someone to do something quicker, easier and cheaper, but might cause widespread unemployment or urban decay. It could also have a harmful impact on the delicate balance of nature.

As you develop your design ideas, you will often need to make important decisions about the social, moral, cultural and environmental impact of your product.

Cultural awareness

People from different cultures think and behave in different ways. What is acceptable to one culture may be confusing or insulting to another. For example, some religions are not allowed to watch TV or use computers. In others only very plain, un-decorated products are acceptable. Colours and certain shapes can have very different meanings across the world.

Environmental issues

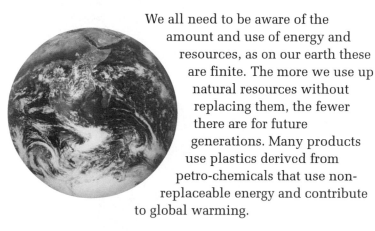

We all need to be aware of the amount and use of energy and resources, as on our earth these are finite. The more we use up natural resources without replacing them, the fewer there are for future generations. Many products use plastics derived from petro-chemicals that use non-replaceable energy and contribute to global warming.

Moral issues

Sometimes designers are asked to develop products that can cause harm to people or animals. Would you be willing to create an electronic product that could hurt someone or could also be used for criminal activity?

Social needs

Good design can help bring people together. Designers need to be careful about creating products that might have the effect of isolating someone, or making them more vulnerable to crime in some way.

Electronic surveillance equipment can be used to listen into conversations and secretly take pictures.

Trevor Baylis's clockwork radio is a good example of socially aware design. It is meant for use in Africa where batteries are extremely expensive. It is made in South Africa by a totally integrated workforce, which includes people with various physical handicaps.

Design Matters (2)

Designing products is a complex task. New designs must be:
▷ *easy and comfortable to use*
▷ *needed and wanted by enough people*
▷ *suitable for manufacture.*

To find out more about ergonomics and anthropometrics go to:
www.usernomics.com/meta.html

The Role of the Designer

One of the key skills of a designer is taking an existing component, such an alarm circuit, working out how it will be housed safely and attractively, and how the various elements can be assembled quickly and easily.

Designers often have to design within considerable constraints. They rarely have a free choice of materials, components and production processes, and often have to work with what a manufacturer already has. Limitations may be imposed by the maximum size a machine can mould, an existing stock of ready-made electronic components and the existing skills of the workforce.

Some products are created with the prime intention of using materials and equipment which are being under-used during the decline in sales of a particular product.

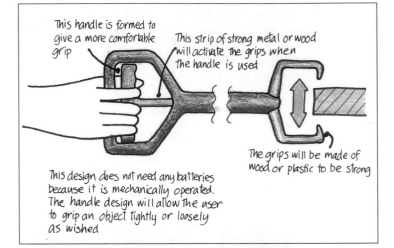

This handle is formed to give a more comfortable grip

This strip of strong metal or wood will activate the grips when the handle is used

The grips will be made of wood or plastic to be strong

This design does not need any batteries because it is mechanically operated. The handle design will allow the user to grip an object tightly or loosely as wished

Ergonomics

Designers need to make sure their products are a pleasure to own and use. This aspect of design is called **ergonomics**, and involves taking into consideration where the product will operate and how it will be used and operated. Things which are small and fiddly to open, or difficult to operate, can benefit from ergonomic study. You need to think carefully about things like:

▷ the weight and size of a product;

▷ the position of displays and controls;

▷ how the product is held and carried, and stored when not in use.

Sometimes this information is already available from books. If not, it may be necessary to set up a series of tests and experiments to obtain the data you need.

Anthropometrics

Anthropometrics is a term often used in connection with ergonomics. It refers to the measurement of the human body; how far we can see, the length of our legs, how much weight we can lift easily, the pressure we can apply with our hand, etc. Statistical data has been collected which gives these measurements as averages for males and females and for all ages from babies to elderly people. This information is important to help you check:

▷ if the intended user will find it easy to physically use the product (e.g. lifting, moving);

▷ whether any unusual skills are required;

▷ what accuracy of observation, decision and response is required;

▷ if instructions, information and control devices are arranged in the best way.

British children	Hand length (mm)	Hand breadth (mm)
3 to 4 years	Boys 100 –125 Girls 100 – 130	Boys 50 – 60 Girls 45 – 60
5 to 7 years	Boys 115 – 145 Girls 110 – 145	Boys 55 –70 Girls 50 – 65

Consumer Demands

Designers need to be clear about the gap or 'niche' in the market that their product is aiming to satisfy, and the sort of people it is aimed at.

People have a wide variety of needs and wants, and are prepared to spend different amounts of money to satisfy their desires. We don't just buy a personal stereo, for example. We want one in a particular price range which will have a specific range of functions, and looks the way we want it to look to reflect our life style. Manufacturers produce a range of models to satisfy different markets. Companies are keen to spot a gap in the market, i.e. a product model or variation which is not well supplied by other manufacturers.

Design Failures

Sometimes products fail to sell in the marketplace. Maybe no-one wanted to buy the product, or there was a better or cheaper alternative. Perhaps it quickly became known that it didn't work well, or was unreliable.

Maybe not enough money was invested in promotion, with the result that not enough people knew it was available, or manufacturing costs proved to be much higher than expected, with the result that no profit was made.

Despite extensive publicity at the time, Sir Clive Sinclair's 'C5' vehicle of the early 1980's failed in the market place.

Industrial Matters

Good design involves creating something that works well and is satisfying to use. But to be successful as a product also needs to be commercially viable.

Technology and the Manufacturing Industry

Improvements in technology during recent years have greatly changed the way products are made. It is now possible to make hundreds of thousands of identical products very cheaply.

Hand-crafted products take a long time to make. The sizes of the different components often vary in size, so each part has to be worked on to ensure everything fits well together.

Parts for mass-produced products, however, can be manufactured to exactly the same size, so all they will all fit one another without adjustment. Using standardised parts, manufacturing can be broken down into a series of assemblies and sub-assemblies.

Control technologies are used throughout the manufacturing process to ensure quality is maintained.

■ ACTIVITY

See if you can find a product in your home which has been hand-made as a one-off. Sketch it and describe how you think it might have been made.

Then find something which has been mass-produced - choose something fairly simple which does not have many parts. Sketch it, and list all its different parts.

Finally explain what you think are some of the advantages and disadvantages of making things by hand and making things by machine. Consider:

▶ quality of manufacture
▶ reliability
▶ appearance
▶ cost
▶ availability
▶ the job-satisfaction of the maker

Personal stereos, portable CD players, minidisks and MP3 players all essentially do the same thing. However a wide range of models are made to cater for different markets. Identical electronic circuits, mechanisms and casings are often used across a range of devices to reduce production costs.

Design for Profit

Products are designed and made to make life easier and more enjoyable, or to make a task or activity more efficient. However, along the way the people who create these products need to make a profit. The designer needs to do more than satisfy the needs of the market, and to consider the sorts of issues described on the previous page. They must also take into account the needs of the clients, manufacturers and retailers. The aim is to design and make products that are successful from everyone's point of view.

Designers:
- agree a brief with a client
- keep a notebook or log of all work done with dates so that time spent can be justified at the end of the project
- check that an identified need is real by examining the market for the product
- keep users' needs in mind at all times
- check existing ideas. Many designers re-style existing products to meet new markets because of changes in fashion, age, environment, materials, new technologies, etc.
- consider social, environmental and moral implications
- consider legal requirements
- set limits to the project to guide its development (design specification)
- produce workable ideas based on a thorough understanding of the brief
- design safe solutions
- suggest materials and production techniques after considering how many products are to be made
- produce working drawings for manufacturers to follow.

Manufacturers need to:
- make a profit on the products produced
- agree and set making limits for the product (manufacturing specification)
- develop marketing strategies
- understand and use appropriate production systems
- reduce parts and assembly time
- reduce labour and material costs
- apply safe working procedures to make safe products
- test products against specifications before distribution
- produce consistent results (quality assurance) by using quality control procedures
- understand and use product distribution systems
- be aware of legislation and consumer rights
- assume legal responsibility for product problems or failures.

Clients:
- identify a need or opportunity and tell a designer what they want a product to do and who it's for (the brief)
- consider the possible market for the idea
- organise people, time and resources and raise finance for the project.

Retailers:
- need to make a profit on the products sold
- consider the market for the product
- give consumers what they want, when they want it, at an acceptable price
- take account of consumers' legal rights
- take consumer complaints seriously
- continually review new products
- put in place a system to review and replace stock levels.

Consumers/users expect the product to:
- do the job it was designed for
- give pleasure in use
- have aesthetic appeal
- be safe for its purpose
- be of acceptable quality
- last for a reasonable lifetime
- offer value for money.

Using ICT in Systems and Control

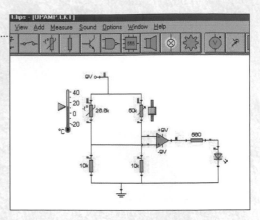

ICT (Information and Communication Technology) is widely used in the design and production of products, as you will discover. You can considerably enhance your GCSE coursework with the effective use of ICT.

Using ICT in your Work

To gain credit for using ICT you need to know when it is best to use a computer to help with your work. Sometimes it is easier to use ICT to help with parts of your coursework than to do it another way. On other occasions it can be far easier to write some notes on a piece of paper than use a computer – this saves you time and helps you to do the job more effectively.

Following are some ideas showing you how using ICT could enhance your coursework. Some can be used at more than one stage. You do not have to use all them!

Identifying the Problem

The **internet** could be used to search manufacturers and retailers web-sites for new products, indicating new product trends.

Project Planning

A time chart can be produced showing the duration of the project and what you hope to achieve at each stage using a **word processor** or **DTP** program. Some programs allow you to produce a Gantt chart.

Investigation

▷ A questionnaire can be produced using a **word processor** or **DTP**. Results from a survey can be presented using a **spreadsheet** as a variety of graphs and charts.

▷ Use a **digital camera** to record visits and existing products.

▷ The **internet** can be used to perform literature searches and communicate with other people around the world via **e-mail**.

Search engines

To help you find the information you need on the internet you can use a search engine. A search engine is a web-site that allows you to type in keywords for a specific subject. It then scans the internet for web sites that match what you are looking for. Here are the addresses of some popular search engines:

www.excite.co.uk
www.yahoo.co.uk
www.netscape.com
www.hotbot.co.uk
www.msn.co.uk
www.searchtheweb.com

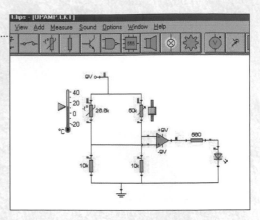

E-mail

E-mail is a fast method of communicating with other people around the world. Text, photographs and computer files can be attached and shared. Some web-sites have e-mail addresses – you could try to contact experts to see if they could help with your coursework. It is important to be as specific as you possibly can, as these experts may be very busy people.

Specification

A design or product specification can be written in a **word processor**. Visual images of the product, diagrams and other illustrations could also be added. Information can be easily modified at a later date.

Developing Ideas

▷ Ideas for your product could be produced using a **graphics** program, **DTP** or **3D design** package. Colour variations can be applied to product drawings to test a design on its intended market before production.

▷ Control systems can be modelled on screen as a flow chart

▷ Electronic circuits can be designed and simulated on screen to check their operation before they are made.

Final Ideas and Production

▷ A document showing the specification, images, production method and components can be **word processed**.

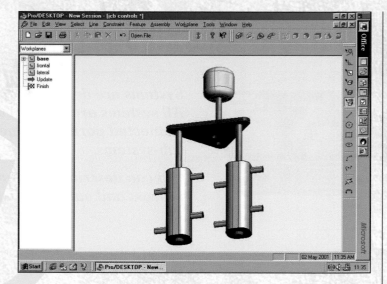

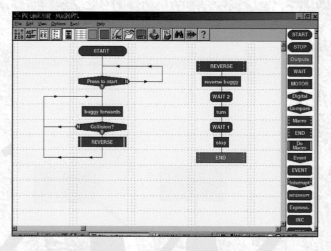

▷ Parts lists and the costs of materials can be calculated and displayed using **spreadsheets**.

▷ A detailed flow production diagram could be produced using a **DTP** program. Images could also be added to show important stages.

▷ **Digital images** can be used in the production plan as a guide to show how the product should be assembled or to indicate its colour.

▷ **Printed Circuit Board production software** can be used to produce PCB masks. More sophisticated software has an 'Autoroute' facility that will produce a PCB mask directly from a circuit design program.

▷ Pre-programmed **CAM** equipment could be used to replicate manufacture (see page 143).

Project Presentation

▷ Use **graphics** packages to prepare text and visual material for presentation panels. Charts showing numerical data can be quickly produced using a spreadsheet.

▷ Use a **presentation** package, such as *PowerPoint* to communicate the main features of your design.

IN YOUR PROJECT

▶ Use a CAD system to design a personalised logo on your project. Then use a computerised engraving machine to put this on your project.

▶ Use a CAD system to design the circuit for an electronic egg timer. Show how by adjusting component values the time period can be adjusted.

KEY POINTS

● CAD will allow to design and check your designs before you start making.

● CAM will help you make your product; especially if you want several parts the same or have complex shapes.

● Generic software can be used to help present your work and product costs.

All About Systems (1)

Systems are devices that perform specific tasks. All systems are composed of a number of inter-connected parts or elements. These are called sub-systems.

You can describe a system in terms of its input, process and output by using a block diagram.

The word *system* has become popular in our language in recent years. People talk about central heating *systems* and comment about the problems experienced with a new one-way traffic *system* in their town. Our attention is drawn to the damage we are doing to the Earth's eco*system* through our use of fossil fuels. But what does the word *system* really mean?

What is a System?

In simple terms, a **system** is a collection of inter-connected parts, or elements, each of which has to perform a specific task if the system is to work. Any change, no matter how small, in any of these parts or elements, will have an effect on the operation of the system.

We can group systems under a number of different headings:

▷ natural systems
▷ designed systems
▷ abstract systems
▷ human activity systems.

Talking about Systems

When explaining how a system works, or describing the individual parts of the system, we have to make sure that we are using the correct language and terminology.

To help us design and understand systems we need to break them down into a number of smaller units. These are called **sub-systems**. The operation of a system, or its individual sub-systems, can be described by using a block diagram.

| Input | → | Process | → | Output |

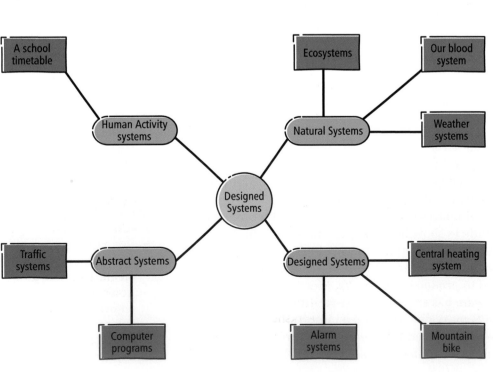

Sub-systems

Most systems are made up of a number of smaller sub-systems, each of which is designed to carry out a particular function. When all of the sub-systems are connected together they form the main system – but remember, they must all be operating correctly otherwise the whole system will not work.

In this book we will concentrate on designed systems, so let's consider one in a little more detail. Remember that a system does not always have to involve electronics, as you will discover later on.

System	Sub-systems	Purpose
House alarm	Sensors	To detect when an intruder is present. Doors are usually fitted with magnetic 'proximity' switches, while sensors in rooms will react to heat and movement.
	Time delay	To allow the user to set, and re-set, the alarm without it going off.
	Operating circuit	Once the circuit has been set the operating circuit waits for a signal from the sensors in order to trigger the sounder. Once triggered, the circuit needs to keep the sounder on. This is known as circuit latching.
	Sounder	All alarm circuits require a sounder that is loud enough to attract the attention of neighbours or the police.
	Wiring and casing	To connect the components and ensure they are held securely.

Block diagrams

When describing how a system or sub-system operates we need to keep the description as simple and clear as possible. In order to do this we use block diagrams. A **block diagram** breaks down the system into three stages. These are:

▷ input
▷ process
▷ output.

This collects the information that will cause the system to trigger.

This is the link in the system that decides when the output needs to be operated.

This is what we usually see or hear when the system has been triggered.

Input
Door and window sensors react to being opened.
In the rooms, sensors react to infrared heat and movement. In order to operate, the alarm must be armed by the user.

Process
Once the alarm has been armed the process circuit waits for one of the sensors to send it a signal.
Once triggered the circuit latches and, at the same time, turns on the output.

Output
The output will sound as soon as a signal is received from the process circuit.

The example above is a block diagram which describes a house alarm system.

■ ACTIVITIES

1. List the sub-systems for the following designed systems:
▶ a television
▶ an automatic door
▶ a remote-controlled car.

2. Use block diagrams to describe the following systems:
▶ a refrigerator
▶ a workshop vice
▶ an automatic garage door.

KEY POINTS

● The word system is used to describe a device that performs a specific task.
● All systems are composed of a number of inter-connected parts or elements. These are called sub-systems.
● We can describe a system by using a block diagram.
● Block diagrams explain a system in terms of its input, process and output.

All About Systems (2)

There are three basic types of system: automatic, semi-automatic and manual.

Flow-diagrams are used to describe the way systems operate.

Types of Systems

Systems come in many different shapes and sizes, performing a variety of different tasks. The requirements of the task will decide whether or not an automatic system, semi-automatic, or a manual system is needed. In the case of a large system that has a number of smaller sub-systems, it may be necessary to have a combination of these.

Automatic Systems

Once it has been started, an **automatic system** is designed to operate without the intervention or involvement of a human user. Automatic systems are usually required to perform repetitive tasks, the operation of which are triggered by a remote sensor, or one that has a fixed cycle. A remote sensor is capable of operating without any physical or manual operation being required. A cycle is one complete operation of the system.

The following are examples of such automatic systems:

▷ a traffic light system – fixed cycle operation
▷ a domestic central heating system – operation controlled by remote sensors.

Systems such as these can be overridden – that is, the user can alter the system in an emergency or if a particular need arises. This should be seen as a safety device, not part of the normal operation of the system.

Traffic light system

We see traffic light systems every day and probably take them for granted. Usually our only comment will be to complain because they always seem to be red when we want them to be green! To describe how this system works we can use a simple flow diagram.

The flow diagram below illustrates the traffic light for one cycle of its operation. This diagram shows the operation of only one set of lights. If an accident is to be avoided, then for every green light there must be another traffic light showing red!

The details of the timing circuits and the light-switching circuits would be described using block diagrams if we wanted to look at those parts of the system in more detail.

This system can also be described as a **closed loop system**, because the whole of the operation of the system is a continuous and repeating cycle.

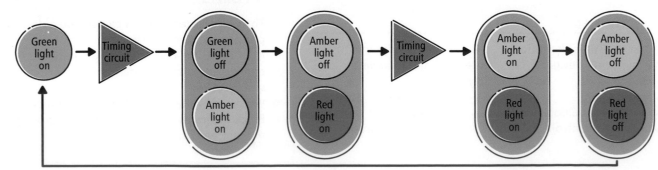

Central heating system

The aim of a central heating system is to keep a house warm and supplied with hot water when needed by the householder. You can find out more about central heating systems on pages 40 and 41.

If we want to describe the operation of this system then the simple flow diagram used for the traffic lights will not be enough. Consider the following flow chart for the heating part only of the system.

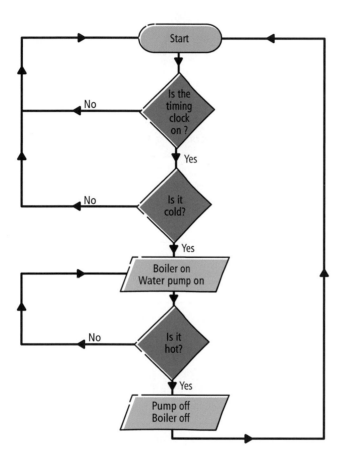

This time the system needs to make decisions as it runs through its cycle of operation. A room thermostat would be used to set the temperature levels required and the householder would decide when the system should come on and go off.

Remember that working alongside the heating system would be the hot water system. The room thermostat and the water temperature sensor both act as the remote sensors needed to trigger the system.

Semi-automatic Systems

In **semi-automatic systems** the majority of the system operation is usually carried out automatically, but the system needs to be triggered by the user in order for it to operate. In our everyday experiences two of the most common semi-automatic systems are:

▷ a lift in a multi-storey car park or shopping centre
▷ automatic shop doors.

Lift system

In order for a lift to operate there are manual and automatic operations that have to take place. Usually the manual operation will set into motion a series of automatic actions before the next manual input is required. Look at the table below:

Manual operation	Automatic sequence
Press button to call lift	– doors close – motor on, brake off – arrive at floor where called – brake on, motor off – doors open.
Press button to send to required floor	– doors close – motor on, brake off – arrive at floor requested – brake on, motor off – doors open

These are only two of the many complex operations that a lift is capable of. Users also have the ability to override the 'close door' command if they want to hold the door open. There are also many safety aspects, including doors that automatically re-open if they sense that something is blocking the doorway. Most modern lifts are also capable of storing memory so that several passengers can select their floors as they enter the lift, and the lift will stop at these floors in sequence.

All About Systems (3)

USER TYPE

Individual   Group   with buggy/pram

Automatic shop doors

They spin, they slide, they open in, they open out: automatic doors, it seems, can be made to operate in a variety of different ways. The rotating door is really an automatic system, as it will continue to operate without stopping except when a blockage in the door is sensed. The others, however, will operate only when a sensor picks up on the approach of an individual. Sometimes the sensors, because of their positioning, will be 'kind enough' to open a door even if the person sensed didn't want to go into the shop!

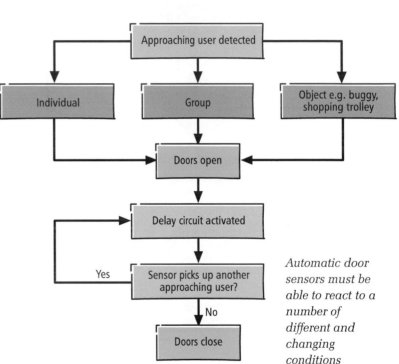

Automatic door sensors must be able to react to a number of different and changing conditions

When setting up the system, the designer needs to make sure that the sensor will detect the approaching potential user and be able to open the door before they reach it, so the user will not have to wait or slow down.

The door mechanism then needs to hold the door open until the user has safely passed through, whoever the user might be, young or old, perhaps pushing a pram or in a wheelchair. The door must not be allowed to close until the user has safely passed through. The range of the sensor is even more important where the door opens towards the approaching user: in this situation the door needs to open while the user is further away.

Manual Systems

Manual systems are often associated with devices that allow the user to make decisions for themselves, with the system being designed to carry out the instructions of the user. Home entertainment systems provide us with a range of examples of manual systems:

▷ televisions
▷ video cassette recorders
▷ stereo systems

Sometimes they are capable of carrying out semi-automated operations, however. These are only activated after a manual input from the system user. Let us consider one of these manual systems in a little more detail.

Video cassette recorder

In very simple terms a VCR is expected to

▷ allow the user to record television programmes.
▷ allow the user to play back from pre- and self-recorded tapes.

The inner workings of the video recorder are very complex. Considering why we can classify the VCR as a manual system? Think about what has to be done in order to programme a VCR to record a programme. The best way to describe the operation is with the use of a simple flow chart.

If the sequence of programming has been carried out correctly you will not miss your favourite programme! However, with manual systems there is a greater chance of an error or a mistake being made, the reason being that they rely upon human input.

A calculator will only give you the answer in response to the keys pressed – the system cannot think for itself!

■ ACTIVITY

Describe the operation of the following systems by means of flow diagrams, indicate where you think there may be a case of feedback:

▶ From pressing a lift call button until you exit from the lift at the floor of your choice.
▶ A set of automated doors that open towards you as you approach them.
▶ The sequence of operations required when loading a new film into a camera.

Using a Peripheral Interface Controller (PIC)

A PIC is a microchip that can be programmed to control a sequence of operations and react to input sensors. They can be used to reliably control complex systems. PICs are widely used in our daily lives, such as programming a video recorder. They are like a small computer, all in one economical small chip. The PICs program can be erased and a new one downloaded. The program can be completed and simulated on screen before it is used.

Using a PIC in your project

PICs are ideal for GCSE coursework projects, such as house alarm systems, complex games, automated vehicles and visual displays. There are many different PICs available. They vary by having different numbers and types of INPUT and OUTPUT pins. However one of the most a useful is the 16F84 (see page 100). This has eight output and five input pins and could be used in a variety of projects.

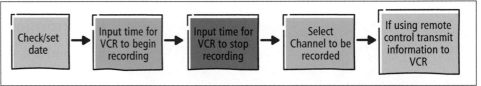

| Check/set date | → | Input time for VCR to begin recording | → | Input time for VCR to stop recording | → | Select Channel to be recorded | → | If using remote control transmit information to VCR |

The sequence shown above may not be the same for every VCR, as different products need to be operated in different ways.

KEY POINTS

● Systems can be divide into three types based upon the way they operate. These are automated systems, semi-automated systems and manual systems.
● System operations can be described with the use of flow diagrams.

System Breakdown (1): The Food Whisk

Many kitchen activities can be carried out with greater efficiency by the use of mechanical and electrical devices. The introduction of the mechanical food whisk with its hand-operated gear system meant that ingredients could be mixed more quickly.

Sub-systems

- **Frame:** allows the user to hold the whisk as well as keeping the mechanism in place.

- **Gear System:** allows the user to rotate the whisk at a high speed with minimum effort.

The food whisk is a simple hand-operated **mechanical system**. Although there are many different types, food whisks all share the same basic design.

The main function of the food whisk is to enable the user to mix together the ingredients of a recipe. Originally a fork was used to perform this task, but this method was often slow and required a lot of energy.

The geared whisk greatly speeded up the mixing process and also required less energy from the user. Continued developments have seen the introduction of electric whisks and food processors, which not only mix but are also capable of chopping food.

Materials

The materials for the whisk need to be chosen carefully because of their use with food. The choice facing the designer is one of cost, whether to choose a material that will not corrode, such as stainless steel, or to use cheaper mild steel, which can then be chrome-plated.

When designing products such as a food whisk or processor it is important to consider ergonomics and anthropometrics.

Ergonomics

Ergonomics is the scientific study of the way in which people use products in their living and working environments. An ergonomic study involves looking at how people perform particular tasks with the aid of their physical capabilities and the use of their senses: sight, sound, smell, touch and taste.

Anthropometrics

Anthropometrics is the statistical study of the human form. Information is collected and recorded for female and male children and adults of different age groups. Just about every human dimension is recorded, including height, hand spans, finger thicknesses, sitting positions, angle of vision, etc. The designer can use this information to help ensure a product will be the right size and shape for the majority of people likely to be using it.

Input	Process	Output
The user rotates the crank handle	The gear system changes the direction of rotation through 90° and increases the speed of rotation	The whisk blades rotate

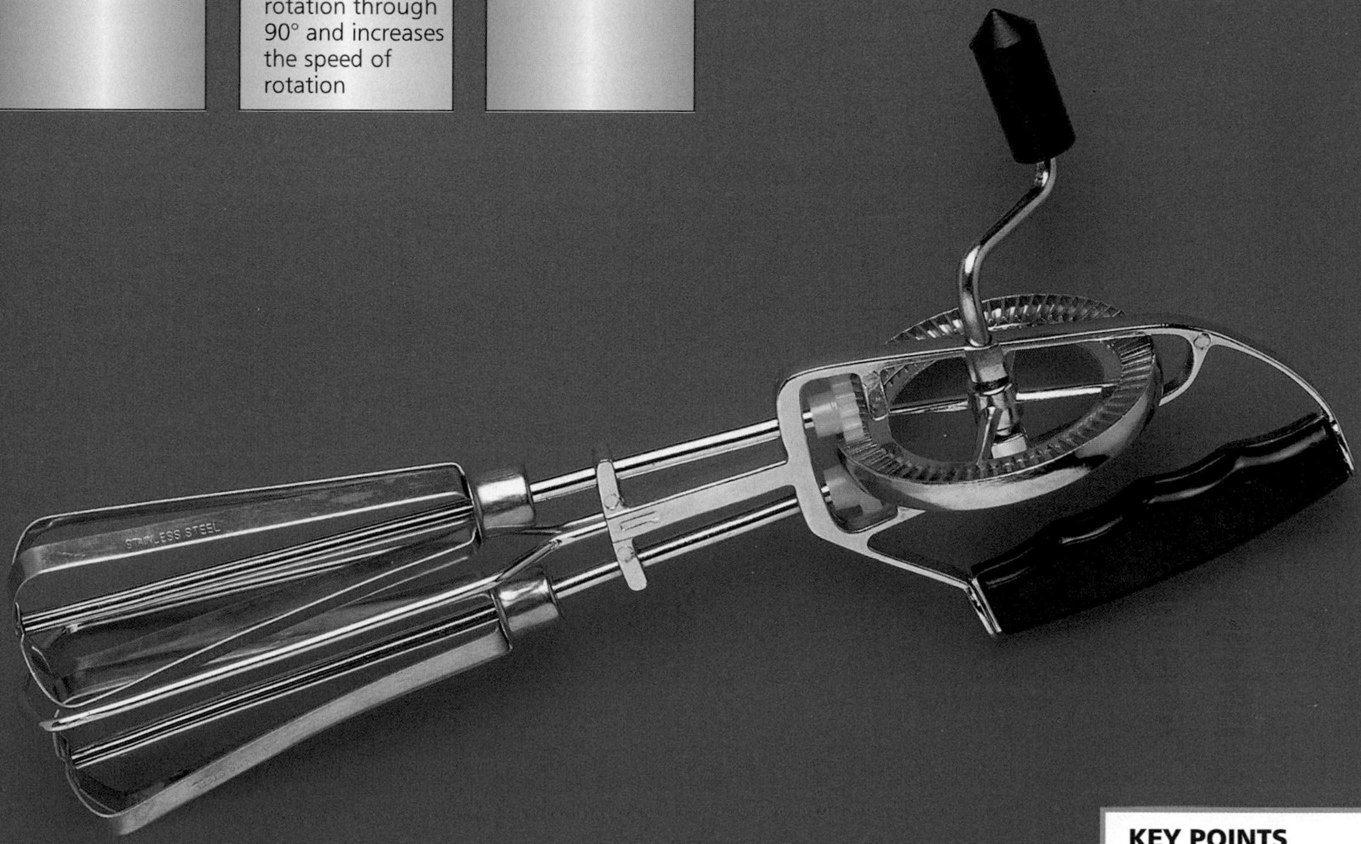

KEY POINTS

- The hand whisk is a mechanical system.
- A gear system that changes the direction of rotation is needed.
- We can calculate the output speed of the whisk from the rate of rotation and the number of teeth on both gears.
- Ergonomics is the study of people in relation to their working environment.
- Anthropometrics is the data that concerns the dimensions of the human body.

Energies
Energy is transferred from the user to operate the whisk.

Associated technologies
- material properties
- gear systems
- anthropometrics and ergonomics.

■ ACTIVITY

Find a hand whisk similar to the one shown in the picture, and find out the following:

► What materials are used to make the whisk? Why do you think these were chosen?
► Calculate the ratio of the gear system used by your whisk. At what r.p.m. would the handle have to be rotated if the whisk blades were needed to rotate at 600 r.p.m.?
► What features of the whisk have been ergonomically designed?

Calculating output speed
In a simple gear system the output speed can be calculated by using a simple formula and some basic information.

Information needed:

► input speed
► number of teeth on **driver** gear – input
► number of teeth on the **driven** gear – output

Formula:
$$\frac{\text{output}}{\text{(r.p.m.)}} = \frac{\text{rotations}}{\text{(r.p.m.)}} \times \frac{\text{teeth on input}}{\text{teeth on output}}$$

System Breakdown (2): A Hair Dryer

The simple task of using a hair dryer is only possible because a designer has succeeded in developing a product which is the right size and shape, efficient to use, safe and inexpensive to manufacture in quantity.

Hand-held portable hair dryers can be found in nearly every home in this country. They have become an almost essential item for many people.

The small miniature models available today have come a long way from those that first went on sale years ago. You can buy hair dryers of many different body styles to suit the many needs and desires of the potential users.

Surface designs have also been added to appeal to potential user groups, from fashion-conscious teenagers to middle-aged mums and dads.

Whatever the shape and design of the outside, nearly all hair dryers operate in the same way, as shown in the system breakdown on the right. In simple terms, they are little more than miniature fan heaters!

Sub-systems

- **Casing:** this holds all internal components in place, as well as allowing the user to hold the dryer.

- **Heating element:** all hair dryers provide warm air, achieved by electricity heating up a wire coil.

- **Motor and fan:** together these draw in cold air, from the back of the dryer, which is then passed over the heating element and pushed out as warm air through the nozzle.

- **Control:** most dryers allow the user to have some control over the speed of the motor and the level of heat required, in a variety of different combinations.

Materials

All user-handled parts of a dryer must be insulators so that the potential dangers of electricity and moisture are kept to a minimum. To achieve this, dryer bodies are usually made from a thermoset plastic such as ABS.

WWW.

To find out more about the history of domestic products go to the Science Museum at:
www.nmsi.ac.uk/welcome.html

Input		Process		Output
Control system: – heat level chosen – motor speed selected – on switch activated.	→	Heating element on. Motor and fan rotate.	→	Warm air streams from dryer nozzle.

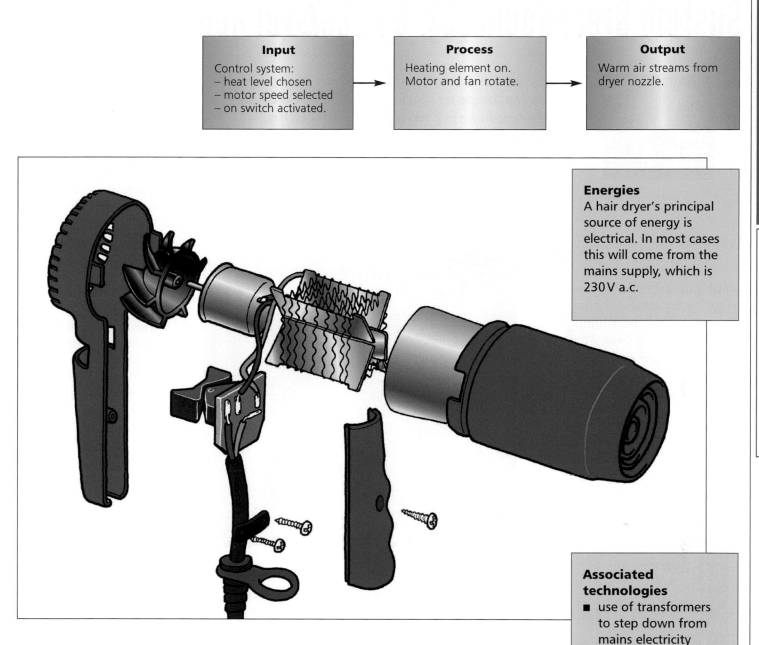

Energies

A hair dryer's principal source of energy is electrical. In most cases this will come from the mains supply, which is 230 V a.c.

Associated technologies

- use of transformers to step down from mains electricity
- working safely with mains electricity
- injection moulding of plastic casings
- a.c. and d.c. motors.

KEY POINTS

- All electrical devices must insulate the user from possible electric shock.

■ **ACTIVITY**

Working in a small group, study a range of different hair dryers and analyse them under the following headings:

▶ Control settings: what settings are available, and how many possible combinations are there?

▶ Switch types: what different forms of switches have been used? Can you name and describe them?

▶ Casing designs: consider the designs and decoration of the casings. What conclusions can you draw from them?

System Breakdown (3): Gas Central Heating

One of the most familiar systems found in most houses keeps us warm and supplied with hot water whenever we need it. The central heating system can be controlled by the user, and reacts to the changing atmospheric conditions within the house.

Until quite recently the prospect of getting up on a cold winter's morning in a house without heating, until the open fire had been lit, was a reality for most people.

Today the story is very different. Most of us live in a house or flat that is centrally heated, using one of the many systems now available.

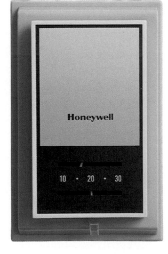

The main function of any central heating system is to maintain a desired minimum temperature inside the house, whatever the temperature outside may be. In order to achieve this, central heating systems bring together a number of sub-systems; this is known as an **integrated system**.

Sub-systems

- **Boiler:** as the name suggests, the boiler heats water for both the domestic water supply and the central heating circuit.

- **Pump, pipework and radiators:** the pump keeps the water moving through the radiator system, returning the cooling water back to the boiler for re-heating.

- **System control:** some form of control over the central heating system is essential to keep temperatures at comfortable levels. The simplest form is a timer control together with a temperature control.

  The advent of microprocessor control systems means that there is an increasing use of this technology in newly installed systems.

Input	Process	Output
Control system: – timer switched on – temperature room is cold.	Gas burner on. Water heats up. Pump on.	Radiators heat up. Room temperature begins to rise.

SYSTEMS AND CONTROL

products and applications

Materials

The principal material used in a central heating system is copper piping. Although quite expensive, copper will not corrode when in contact with water.

Systems feedback

An efficient central heating system takes account of the changing conditions within the home. If it is cold then the heating needs to come on, and as the temperature rises to the level set it then needs to turn itself off. The system then continues to monotor the situation, turning the heating on and off as necessary.

The term we use to describe these actions is **feedback**, where the system uses information to react to changing conditions. In the case of the central heating system these changes are made automatically and this is known as a **closed loop feedback**.

WWW.

To find out more about how everyday things work, go to:
www.howstuffworks.com/ index.htm
www.http//home.clara.net/ austin/thunkin.html

Energies

Thermal energy is transferred from natural gas as it burns, in order to heat up the water.

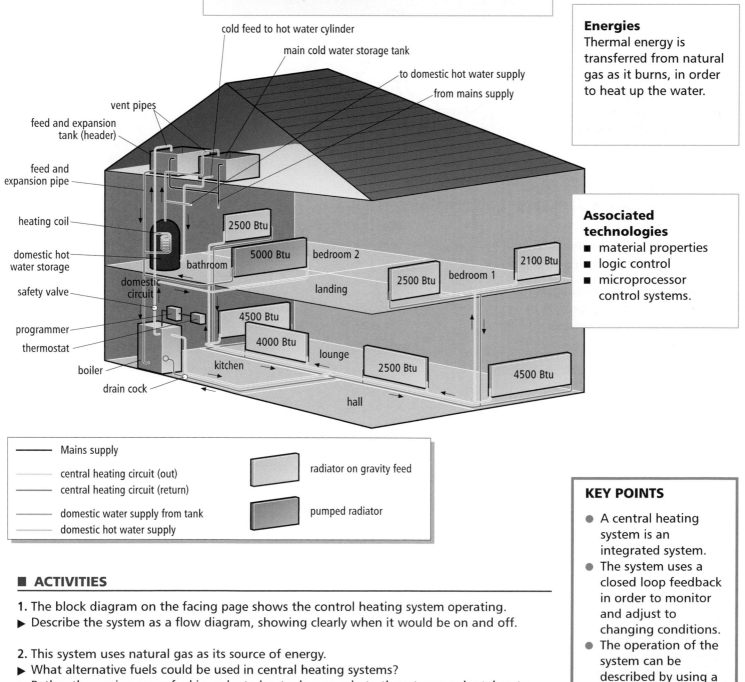

cold feed to hot water cylinder
main cold water storage tank
to domestic hot water supply
from mains supply
vent pipes
feed and expansion tank (header)
feed and expansion pipe
heating coil
domestic hot water storage
safety valve
programmer
thermostat
boiler
drain cock
domestic circuit
2500 Btu
5000 Btu   bedroom 2
2500 Btu   bedroom 1
2100 Btu
landing
4500 Btu
4000 Btu
lounge
2500 Btu
4500 Btu
kitchen
hall
bathroom

Associated technologies

- material properties
- logic control
- microprocessor control systems.

Mains supply
central heating circuit (out)
central heating circuit (return)
domestic water supply from tank
domestic hot water supply
radiator on gravity feed
pumped radiator

KEY POINTS

- A central heating system is an integrated system.
- The system uses a closed loop feedback in order to monitor and adjust to changing conditions.
- The operation of the system can be described by using a flow diagram.

■ ACTIVITIES

1. The block diagram on the facing page shows the control heating system operating.
▶ Describe the system as a flow diagram, showing clearly when it would be on and off.

2. This system uses natural gas as its source of energy.
▶ What alternative fuels could be used in central heating systems?
▶ Rather than using more fuel in order to heat a house, what other steps can be taken to ensure that any heat generated is not lost?

System Breakdown (4): The Mountain Bike

Cycling is an increasingly popular pastime. One reason for this is the development of the modern mountain bike, with its strong, light, frame construction, and its healthy outdoor image.

The basic design of a bicycle has not really changed since John Kemp Starley designed the Rover Safety Cycle in 1880. The modern mountain bike uses the same basic principles:

▷ a diamond-shaped frame made from steel tubing

▷ two similarly sized wheels

▷ the upright riding position.

However, the performance and durability of bicycles has greatly improved through continual product development and the use of modern materials.

The bicycle is the most energy efficient form of transport, especially when you consider most journeys we carry out are under five miles. It can also carry quite large loads, over reasonable distances. When used for journeys around busy towns and city centres it is often the fastest form of transport too.

The bicycle will need to provide a central role in meeting our transport needs and environmental targets for the 21st century, if we are to overcome the traffic jams and pollution we now experience.

Sub-systems

■ **Frame:** this supports the rider, wheels and transmission. It needs to be strong and able to absorb sudden shock loading when going over rough ground.

■ **Gears:** the gear ratio is changed by using different sized sprockets on the rear wheel.

■ **Brakes:** these use levers to increase the force applied by the user.

■ **Speedometer:** a small magnet fixed to the front wheel operates a reed switch, which is connected to the speedometer. Digital electronics are used to calculate and display the bicycle's speed.

■ **Transmission:** the sprocket and chain efficiently transfer the rotary motion of the crank and pedals to the rear wheel.

Materials

Bicycle performance has improved rapidly with the use of modern alloys and composite materials. These need to be light, strong and weather resistant, like carbon fibre. Once finished, their appearance is also important, as this will help sell the bike.

Input	**Process**	**Ouput**
User rotates cranks.	Chain and sprocket transfer rotary motion.	Bicycle moves along.
User pulls brake lever.	Lever increases force and cable transfers the motion.	Brake blocks grip wheel to slow bicycle.
Magnet on front wheel sends pulse to speedometer.	Speedometer uses an integrated circuit (IC) to count pulses and work out speed.	Liquid crystal display (LCD) shows the speed.

WWW. ➡

To find out more about bicycles and cycling go to:
www.exploratorium.edu/ cycling
www.raleigh.co.uk

SYSTEMS AND CONTROL

products and applications

Energies

The bicycle transfers energy from the user in order to make it move. Our bodies can transfer the energy in food we eat to provide the force to turn the pedals on the bicycle.

The brakes use friction to slow the bicycle down.

The force we use to operate the levers is transferred to the brakes by cables. The cable then pulls the brakes on.

Associated technologies

- material properties
- anthropometrics and ergonomics
- energy conversion
- social implications.

■ ACTIVITIES

1. Calculate the gear ratio of the chain and sprocket system on a bicycle. You can find out how to do this by reading page 133.

2. Calculate the mechanical advantage (force multiplication) of the brake lever on a bike. You can find out how to do this on page 133.

3. Draw arrows to represent the forces on a bicycle frame when someone is riding along a straight road.

4. How could you modify a bicycle speedometer to record how far a jogger runs?

KEY POINTS

- The bicycle uses mechanical, structural and electronic systems.
- Modern materials and their properties have improved the quality of new bicycles.
- The bicycle forms part of our transport system for the 21st century.

System Breakdown (5): The Pager

Pagers enables us to send urgent messages to other people who are not near a telephone. They use electronic and communication systems to send and display information.

The pager is a popular product that makes use of the very latest communication and information technology. The user may be alerted that they are needed and can return the call by using a telephone. A message may also be displayed on the pager, so the user may know why they have been contacted.

The message is sent to the pager by using technology similar to a radio transmitter. This message can then be displayed on a liquid crystal display (LCD). The pager can be made to sound an alarm or vibrate, to alert the user that a message has been received.

Pagers are regularly used by people such as doctors, repair or maintenance technicians and business people. The modern pager is a very small and discreet product, that will fit in a pocket.

Sub-systems

- **Liquid crystal display:** LCDs are widely used in many electronic products, such as calculators and mobile phones. They use very small amounts of electrical energy to run. This means the battery will last a long time.

- **Selector switch:** this will control the pager's different features, e.g. displaying the message and turning the audible alarm off.

- **Belt clip:** this will be made from spring steel or moulded plastic and enables the pager to be carried safely on the user's belt.

- **Audible alarm:** a small buzzer will alert the user that a message has been received. It will be driven by an astable circuit, so it sounds intermittently.

- **Vibrating alarm:** a small electromagnetic device will make the pager vibrate to alert the user. This type of technology can be used in alarms for people with poor hearing. They can be placed under the user's pillow, to wake them in the morning, or can be connected to a fire alarm.

Person sending message telephones central control

Central control receives message and transmits it

LUNCH AT 12:30

Person receives message

26/08/97 13:01

Input		**Process**		**Output**
Radio sgnal received.	→	Converts signal.	→	Audible alarm or vibrating and display.

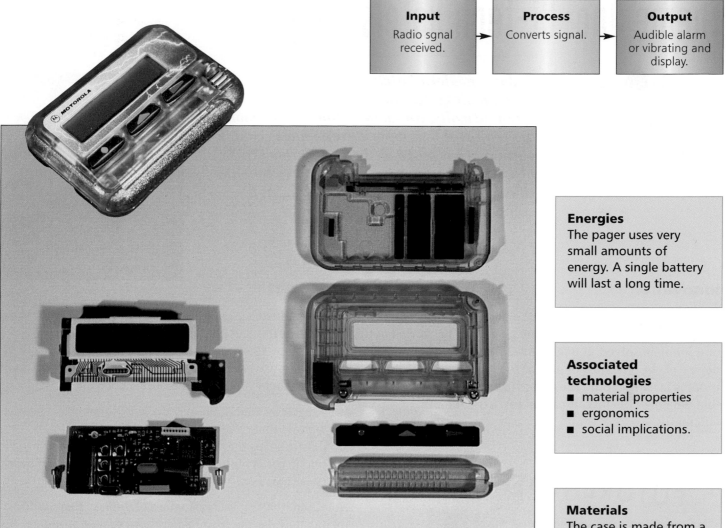

Energies

The pager uses very small amounts of energy. A single battery will last a long time.

Associated technologies

- material properties
- ergonomics
- social implications.

Materials

The case is made from a moulded plastic such as ABS. This can be injection moulded into a variety of shapes, which can then be assembled easily with the other components, such as the circuit board and LCD.

It's4U

The recent growth of mobile phones that can send and receive text messages has prompted the demise of the pager.

■ ACTIVITIES

1. Explain the main ergonomic considerations in the design of a pager or a mobile phone.

2. Explain why a user may wish to turn the audible ringing tone off, or change the way it sounds.

3. Produce an initial specification for a moulded case for a pager or mobile phone aimed at the teenage market.

Is this what a phone of the future will be like?

KEY POINTS

- Pagers are widely used by people who need to be contacted in an emergency.
- They use the latest miniaturised electronics and information technology.

Theme One: Hi-tech Hotels

Today's modern hotels are places full of hi-technology equipment. Doors open automatically, air conditioning keeps rooms at comfortable temperatures, fitness centre equipment is state-of-the-art. Hotels are always looking for innovative ways to make their facilities that bit better than their competitors'.

Innovation

The art of innovation involves the designer thinking up new and different ways in which a task can be carried out or a problem solved. It might involve the invention of a device or the changing of a way in which a particular process is carried out. The aim is to try and make sure that whatever the task, it can be carried out more efficiently.

FUTURE HOTELS

Future Hotels wish to promote our up-to-date use of technology in our newest construction 'The Millennium Hotel'. We are, therefore, looking for exciting new ideas for making our guests' stay more comfortable and enjoyable and, at the same time, improve the working conditions and efficiency of our staff.

The purpose of my writing is to ask you to submit for consideration your ideas of how we might 'smarten up' the Millennium in the areas of:

- automatic control systems
- improving temperature control in the kitchens
- the fitness centre.

We look forward to receiving your submissions.

Yours faithfully

Chairman
Future Hotels PLC

Get Smart!

Smart products and systems use materials that can change their properties and characteristics in response to the environment. Designers can then use these in a variety of ways to allow devices to operate remotely. There are many smart products already available, such as:

▷ sunglasses that darken as the sun shines
▷ T-shirts that change colour because of heat
▷ plastic strips that test the energy of a battery
▷ phone cards of different values that are updated every time they are used.

Hotels are making increasing use of devices that could be included under the list of smart devices and products.

Key facts

The use of keys in hotels is becoming a thing of the past, with guests using credit-card-like 'keys' that are coded for use only on one particular door. This allows the hotel to offer a greater standard of security, as a room code can easily be changed so that the 'key' can be used only for the duration of the visitor's stay.

Making sense

A wide variety of sensors are used in a hotel – for example, to turn lights on and off when guests enter and leave a certain area, or turn off a light that has been left on.

Lift systems are full of sensors and input devices such as user-operated buttons and other sensors that are automatically activated. Till systems and booking systems also make use of smart technology.

Some Like It Hot

Energy is easy to waste and hard to control. At home, in winter, it can be quite difficult to get the temperature just right. The fire is on, but it's often just a bit too warm and stuffy, or a little too cold and draughty – so you open the window, or put on a warmer jersey. Getting the temperature right in a large hotel is a lot more difficult.

Finding the cheapest and most effective way to heat and cool a hotel is a major task. A modern energy management system can help to overcome the problem of supplying different amounts of heat to different parts of a

hotel at different times. Sensors can monitor the temperature at different locations, and the moment the temperature changes, a computer-controlled system can switch on or cut out fans, pumps or radiators as necessary. The system can store and print out a record of what has happened, so that the data can be analysed and problems fixed.

An efficient system can help provide a more comfortable atmosphere for staff and guests, and save energy.

Case Study

Ove Arup and Partners were commissioned to design a co-ordinated energy centre for the Birmingham Convention Centre, Sheraton Hotel and National Indoor Arena development scheme. The three buildings are served by one central energy plant, and the centre can be monitored and controlled at one computer terminal.

Combined with high-performance insulation of pipes and boilers, the system provides a highly efficient and controllable supply of energy.

The same computer program is also used to control the electrical, lighting, fire-alarm and security systems.

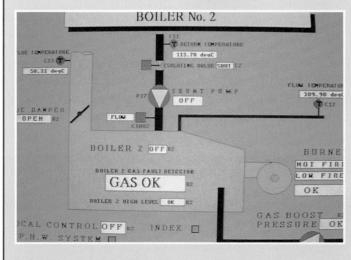

The on-screen information enables the operator to check the performance of every boiler, valve, pump and fan in the system. The computer is programmed to make the most efficient use of the energy system with each zone of the buildings. Detailed warning can be given of any faults which are developing and it is possible to review the state of the system at some previous time.

Project One: Starting Point

Guests entering and leaving large hotels are often carrying cases and bags. As a result they do not have their hands free to open and close the hotel doors. In order to solve this problem many hotels now use automatic doors similar to those found in large supermarkets.

Sensors
(pages 50 and 52)

Safety First
(page 68)

System Modelling
(pages 64 and 66)

Preparing for the Task

Read through the task on the right. Before you start to develop and finalise your suggestions for a design solution you will need to work through the sections which make up this project.

First you will need to learn more about how sensors work, and how they need to be programmed using logic. Next you will look at pneumatic systems to understand how they work. Finally you need to learn more about ways in which systems can be modelled using kits.

The Task

Welcome to the Millennium Hotel

It is important that guests feel welcome when they arrive at the Millennium Hotel. Automatically opening doors will allow guests to enter and exit the building without unnecessary delay or difficulty. Below is a general outline specification for the doors:

▷ The doors need to be sliding or opening.
▷ The doors need to open in advance of the guest.
▷ The doors must stay open long enough to allow the guest to enter safely.
▷ The doors must be able to sense the approach of a second guest even if they are already open.
▷ Hotel staff should be able to override the system for reasons of safety.

The Hotel anticipates that the doors will be of a size and construction that makes them both larger and heavier than normal. The opening mechanism therefore needs to be powerful enough to cope with this situation.

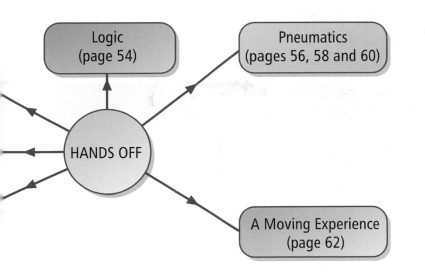

Starting Points

Before you start to develop your ideas you will need to consider the following.

Investigation

What types of automatic doors are already in use? A good source of information will be shopping centres and supermarkets.

Describe the operation of the different doors by means of flow diagrams. What safety features have been built into their operation? Compare the ways in which the doors open. Does this have an influence on the operation of the system? You will also need to make notes and diagrams under the headings on the right.

System input

How is the operational cycle of the doors activated? Find out all you can about the system by researching into the different inputs that are used.

System output

The final output of the system is that the doors will open/close, but how is this made to happen and, what is the method of power used? What mechanisms have been used for opening and closing?

System process

Given the input and output to the system, what process components can be used in order to make the system operate efficiently and reliably?

System modelling

The project requires that you make a working model to demonstrate your solution to the problem. You need to identify the different sub-systems of the whole system and consider what will be the most appropriate method of modelling to use when presenting your finished product.

■ ACTIVITY

How many lessons have you got to complete this project?

In your project folder make a planning chart which shows the main stages of your project and by when you need to have completed them.

Sensors (1): Triggering the System

Every system needs to be triggered if it is to be made to work. Most systems rely on some type of sensor to act as the input into the system. Its function is to trigger the process/control element.

Putting Sensors in their Place

A sensor usually forms part of the input to a system. In many systems there may be more than one sensor required to, or capable of, triggering the system – for example, a household burglar alarm:

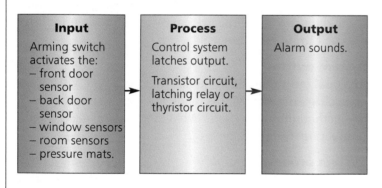

Input	Process	Output
Arming switch activates the: – front door sensor – back door sensor – window sensors – room sensors – pressure mats.	Control system latches output. Transistor circuit, latching relay or thyristor circuit.	Alarm sounds.

Sensors: a Definition

In this book, system input sensors are divided into two groups: **remote sensors** and **physical sensors**.

Remote sensors are capable of operating without any physical or manual operation being required – for example, central heating thermostats and passive infra-red security lighting. As a general rule, remote sensors react to a change in conditions such as light or heat, whereas physical sensors require an external force to be applied in order for them to be activated. Physical sensors act, therefore, as a switch.

In many systems the sensor forms a part of the system input. Physical sensors can often be used to provide a system with valuable feedback.

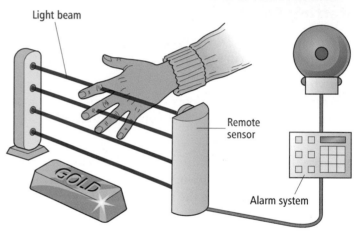

Light beam

Remote sensor

Alarm system

GOLD

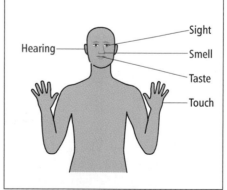

Sight
Hearing
Smell
Taste
Touch

The human body is probably the most complex of all systems. In order for us to operate efficiently many of our processes are triggered by one or more of our five senses.

Which sensor to choose?

Designers need to have a clear understanding of the function of the system and how that system will need to be triggered before they can decide which sensor to choose. The skill lies in the ability to choose the input sensor, or sensors, that will enable the system to operate efficiently and reliably. There are many sensors to choose from.

Remote sensors as input devices

The following components control the process element of a system by using a remote sensor:

▷ temperature dependent resistor
▷ light dependent resistor
▷ moisture detector
▷ reed switch
▷ proximity switches.

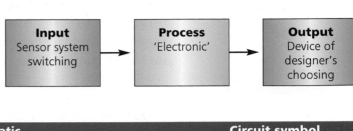

Input	Process	Output
Input Sensor system switching	**Process** 'Electronic'	**Output** Device of designer's choosing

Remote sensors

Condition	Sensor	Schematic	Circuit symbol
Light or dark	Light dependent resistor		
Hot or cold	Temperature dependent resistor (Thermistor)		
Heat	Infrared detector		
Wet or dry	Moisture detector	copper track / PCB board	
Magnetic field	Reed switch Proximity switches		

When used as the input to a circuit these sensor/resistor arrangements will react to the conditions indicated, when placed as shown below.

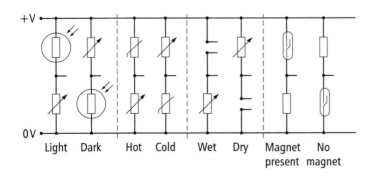

+V

0V

Light   Dark   Hot   Cold   Wet   Dry   Magnet present   No magnet

■ ACTIVITY

Describe what systems could make use of the remote sensors listed above. In each case set out the necessary operating specification that the system needs to follow.

IN YOUR PROJECT

Show, by means of annotated sketches, how a light dependent resistor and/or an Infrared detector could be used for the system you are designing.

KEY POINTS

● Remote sensors are activated by a change in conditions.
● Systems making use of remote sensors are usually automatic or semi-automatic.
● A remote sensor does not need to rely on the direct operation of the system user.
● Remote sensors form the input stage of a systems block diagram.

Sensors (2): Sensing with Impact

Physical sensors are often in the form of a microswitch. They usually need them to be triggered by the system user or by some moving part of the system.

Introducing Physical Sensors

A physical sensor is often placed before a remote sensor in the input stage of a system. If the system needs to have an on/off control then the designer must select an appropriate manual switch. In many systems, however, the operation of the system is used to trigger an input device in another part of the system.

System requirement	Sensor	Application
Input signal to show that a device has successfully operated.	Microswitches are activated when the wheels are in place, indicator light on control panel illuminates.	Aircraft pilots need to know that the wheels of their plane are down before they land.
Input signal to allow motor to operate when certain safety levels have been met.	Microswitch must be closed in order to allow the machine to operate.	The drill in the school workshop will only start when the drive pulley cover is securely closed.
The system will not operate until it is activated by a master control.	A system override switch allows the system to be activated or de-activated when necessary.	The alarm sensors in a house become active only once the master switch has been operated.

The table on the right shows when a physical sensor might be needed in a system, together with an example of a potential application.

Choosing physical sensors

All physical sensors rely on a connection being made, or broken, if they are to operate. All switches operate in this way. What makes one switch different from another is how that connection is made or broken when the switch is in use.

More detailed information on the different types of manual switches can be found in *Design and Make It! Electronic Products.*

With a mouse connected to a computer it is easy to point at anything on the screen. The movement of the ball is translated into left-right and up-down instructions, sent as pulses to the computer

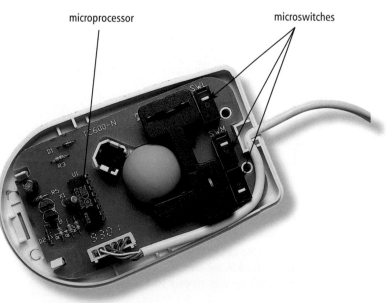

microprocessor

microswitches

Microswitches

Whenever we see the word 'micro' we immediately think of something small. Although microswitches can be small in size, they are called 'micro' because a small force can operate them.

Microswitches are extremely versatile and are used for many different purposes. In order for them to operate efficiently the system designer needs to choose from a range of actuators that can operate the switch. Some possible uses and actuator types are shown on the right.

Actuator type		Microswitch application
Button	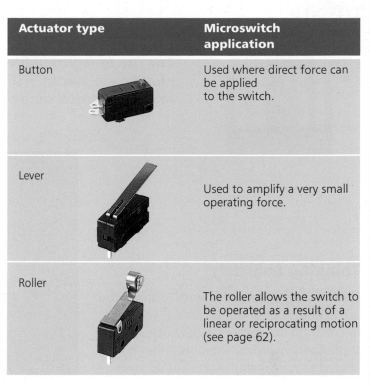	Used where direct force can be applied to the switch.
Lever		Used to amplify a very small operating force.
Roller		The roller allows the switch to be operated as a result of a linear or reciprocating motion (see page 62).

■ ACTIVITIES

1. Microswitches are used in a wide variety of systems. Find five systems that make use of microswitches and briefly describe the purpose of the microswitch in each system.

2. A museum wants to keep a record of how many people entered the building through the entrance in a day. The block diagram for the counting system is shown below.

Input	**Process**	**Output**
Door opened	Microswitch	Counter advances one

IN YOUR PROJECT

Suggest which microswitch trigger would be the most appropriate, given the different types of door that could be used. Make sure that you support your decision with an explanation of how the operation of the door will ensure your chosen actuator will be operated reliably.

ICT ➡

Use a CD-ROM component catalogue to find your required switch.

KEY POINTS

● Physical sensors are usually triggered by the operation of a system part rather than by a human operator.
● Microswitches are the most common form of physical sensor.
● Changing the switch actuator can control the microswitch operation.
● Microswitches can be used to limit the operation of a system by:
  • causing the system to stop working when a limit has been reached
  • not allowing a system to work if safety covers are open.

Logic

The use of logic enables the system designer to plan how a system will operate before having to decide which components will be needed.

Thinking Logically

It's December, it's cold and it's raining. You are in your nice warm house but you need to go out to the corner shop. Looking out through the window you notice that it has started to rain again, so you put on your warm coat and find an umbrella before setting out on your journey.

At one time or another we have all had to face situations like this. In doing so we have been thinking logically. In systems and control technology we can describe the event by means of a **logic diagram**.

Logic gates

In the diagram on the left you will notice the addition of a strange looking box. This is the symbol of a logic gate – in this case an AND gate. **Logic gates** are a convenient way to indicate which inputs are needed in order for a system to operate. To decide which logic gate to use the simplest method is to describe the system.

Let's describe the situation once again.

> *You are about to go out AND it is raining THEN put on your coat and take an umbrella.*

Another example:

> *It is windy OR it is cold AND are you about to go out THEN put on your coat.*

This example introduces a second logic gate, an OR gate. In this situation we are using the output from the OR gate to act as an input to the AND gate. Diagrammatically we would draw this as shown on the left.

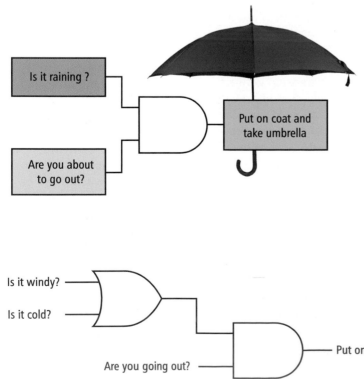

Designing with Logic

In a system, electronic switches, pneumatic and hydraulic valves or IC logic chips (e.g. CMOS 4081B and CMOS 4071B) can carry out the logic gate functions.

To have a better understanding of how a design problem can be developed into a design solution using logic consider the following problem.

Winter mornings often bring with them the problem of frost. Consider the design of a system that will warn the user that the temperature has fallen below 0°C overnight.

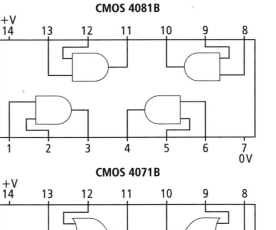

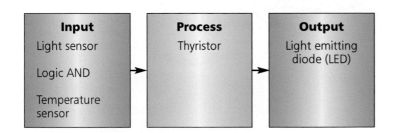

Input	Process	Output
Light sensor	Thyristor	Light emitting diode (LED)
Logic AND		
Temperature sensor		

The circuit can be described with the following control statement:

> *WHEN the light sensor registers that it has gone dark AND the temperature sensor registers that it has fallen below 0°C THEN the LED should switch on.*

A solution to the problem can be demonstrated using an AND gate, as shown on the right. The action of this AND gate can be described by its **truth table** (see page 73).

The AND function is provided by the 4081B logic chip. A design solution is shown below.

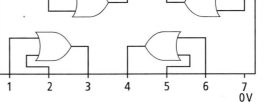

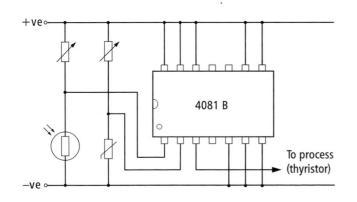

To process (thyristor)

Y	Z	Q
0	0	0
0	1	0
1	0	0
1	1	1

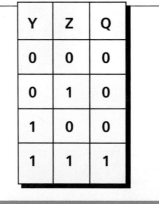

IN YOUR PROJECT

Look again at the specification for your system. Present the specification as a logic diagram.

KEY POINTS

● Logic gates enable us to explain simply the operation of a system where different inputs are involved.

● An OR operation will enable the process to be activated if any of the input conditions are met.

● An AND operation will enable the process to be activated only if all the input conditions are met.

Pneumatics (1): Under Pressure

Pneumatic components and devices use compressed air to make the system operate. As with electronic systems, they can be broken down into input, process and output components.

Pneumatics Systems: an Overview

If you have ever pumped up a tyre on your bicycle then you have operated as part of a pneumatic system. You provided the input to the system by operating the pump, which is the control. The output of the system is the inflating of the tyre.

There are examples of pneumatics all around us:

▷ pneumatic drills digging up our roads
▷ pneumatic hammers to remove and fit the tyres on Formula One cars in record time
▷ air brushes for use in an art lesson.

Most industrial applications do not rely on muscle power to provide the compressed air, but instead use a petrol or electric motor attached to a pump and reservoir.

Creating Pneumatic Systems

In order to build a pneumatic system we need to have a clear understanding of the task that is required of the circuit, together with a knowledge of pneumatic components. The examples shown above are very simple systems that use a minimum of components. In many applications, however, pneumatic systems have to carry out a task that requires a sequence of operations to take place.

Inputs

Every system needs an input, and pneumatic systems are no exception. There are many devices that can be used as an input to a pneumatic system – the key feature being that, whatever the input device, it must cause a valve in the system to be operated.

The pneumatic valve that is operated by the input is either a 3-port or a 5-port valve.

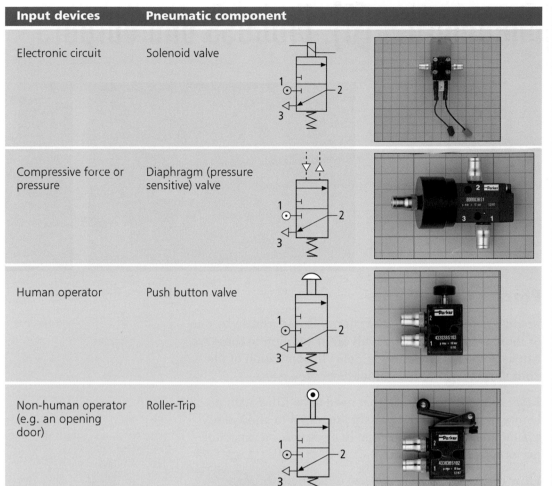

Input devices	Pneumatic component		
Electronic circuit	Solenoid valve		
Compressive force or pressure	Diaphragm (pressure sensitive) valve		
Human operator	Push button valve		
Non-human operator (e.g. an opening door)	Roller-Trip		

Pneumatic Valves Explained

In a human body valves are used to control the flow of blood in the heart, arteries and veins. In a similar way different types of valves are used to regulate the flow of compressed air in a pneumatic system.

A 3-port valve is used to control the flow of air in a circuit. The ports of the valve can be opened or closed by a mechanical, electrical or pneumatic switch. Normally a 3-port valve is used to control the flow of air from the main supply to other components in the circuit.

A 5-port valve has the switching capability of two 3-port valves operating together. Normally 5-port valves are rarely used to turn off the air supply. Instead they act like an OR gate switching the compressed air output between two of its ports. Their most common use is in the control of a double acting cylinder, as explained on page 58.

In situations where an OR gate valve is needed, but direct operation of the valve is not possible, then the most suitable component is a pressure operated 5-port valve. In addition to the five operating ports, the pressure-operated valve has two signal ports which are used to control the operation of the valve.

A 3-port valve

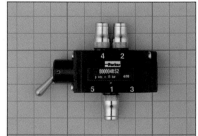

A 5-port valve

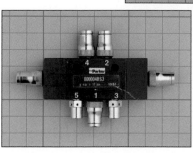

A pilot valve

pneumatics

Pneumatics (2): Process and Outputs

Process Components

In any pneumatic circuit the control is undertaken by one or more 3- and 5-port pilot valves. In addition to these we can use other components to delay the operation of the circuit or to control the output devices.

Reservoirs are containers that need to be filled with air before they achieve the pressure required to allow the system to continue to operate. In electronics a capacitor performs a similar function.

Flow regulators reduce the size of the hole through which the air in the circuit can flow. An adjustable flow regulator performs a role similar to that of a variable resistor in an electronic circuit.

A reservoir

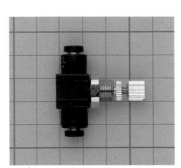

A flow regulator

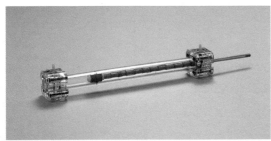

A single acting cylinder

Pneumatic Outputs

Although we can use compressed air to drive a fan and so operate a device such as a dentist's drill, the most commonly used pneumatic output is the **cylinder**. There are two types that are available: the **single acting** cylinder and the **double acting** cylinder.

A double acting cylinder

Putting a System Together

Individual pneumatic components can be combined to form a complete and working system that can perform a pre-determined sequence and repeat this sequence again and again.

Example

A small production company that makes wooden wind chimes wants to automate the drilling of the completed chime blocks. They have decided to use a pneumatic system that will hold and drill the chimes as they are completed.

The sequence of the system is as follows:

1   Completed chimes blocks are placed into the hopper.
2   First chime block is pushed into position ready to be drilled. Note some form of guide system will be needed.
3   When the chime block is in place the drill is lowered to cut the hole and then returns to its starting position.
4   The next chime block falls into place and as this is moved into position it ejects the previously drilled chime block.
5   The system then continues to repeat the sequence.

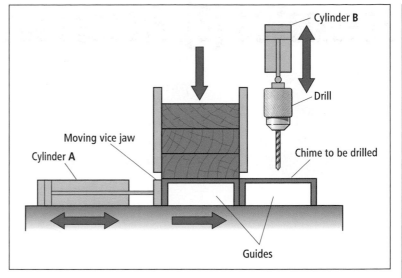

A schematic diagram of the solution

The circuit diagram of the solution

■ ACTIVITIES

1. A sheet metal company uses a pneumatically powered guillotine to cut large sheets into smaller sizes. For reasons of safety, the guillotine will only operate when the following logic diagram has been satisfied.

Suggest which input devices you consider to be the most appropriate for the system.

2. A pneumatic cylinder operates the guillotine blade. If any of the inputs are released then the blade should automatically return to its starting position. Which type of cylinder would you recommend to operate the guillotine blade? Explain your answer.

L hand operator button —
Guard closed —
R hand operator button — → To process

IN YOUR PROJECT

A pneumatic cylinder provides linear or reciprocating movement (see page 62). This is ideal for a sliding door. Show how a pneumatic cylinder could be used to open and close a hinged door.

KEY POINTS

● Pneumatic systems are operated by compressed air.
● As with electronic systems, pneumatic systems can be described in terms of input, process and output.
● Pneumatic system inputs do not have to be pneumatic based.
● Pneumatic valves all operate in the same manner, but the actuators can be changed to react to different input forces and motions.

Pneumatics [3]: Air Control

A pneumatic cylinder can move very quickly. Sometimes it is important to control its speed or delay a cylinder. It is also possible to control the sequence (order) in which a number of cylinders is operated.

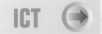

ICT

Build up a library of component pneumatic symbols in your word processor.

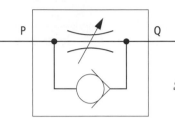

Symbol for a flow regulator

Controlling the Flow

It is possible to control the speed of a pneumatic cylinder by reducing the amount of compressed air that reaches it. This can be achieved by using a flow regulator.

A flow regulator restricts the air flow in only one direction, while in the other direction it is free to flow. The rate at which the air is slowed can be changed by using the adjustment screw on the regulator.

A flow regulator can be used to restrict the flow *going to* the cylinder. It is best used restricting the flow *leaving* a double acting cylinder, as shown on the right.

When the valve is depressed, the air flows and pushes the cylinder. The air trapped behind the moving cylinder tries to escape through the flow regulator, which slows the escaping air. This back pressure in turn slows the movement of the cylinder. A second flow regulator can be added to slow down the cylinder in both directions.

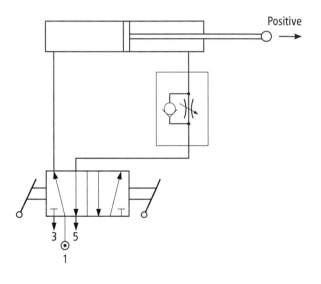

Time Delays

It is possible to delay the operation of a cylinder by using a flow regulator and reservoir connected in series.

When the valve is depressed air flow is restricted by the flow regulator. The reservoir is like an empty bottle that slowly fills up with air. Once it is full the cylinder will start to move. The length of the delay is controlled by the setting on the flow regulator and size of the reservoir.

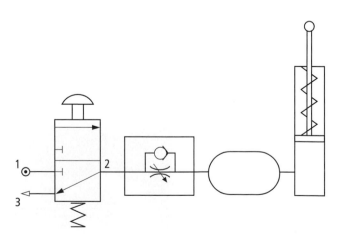

Air Operated Valves

Instead of being switched by a lever, **air operated valves** are controlled by air pressure. This air supply is called a **pilot signal** or **signal line**. It is possible to obtain air operated 3- and 5-port valves. It is important to understand that the low pressure air used to operate the valves is different from the air used to operate the cylinders. This can be used as a safety feature.

It is the same as using a low voltage electronic circuit to control a mains operated device, such as a remote control switching on a television.

Force calculation

It is possible to calculate the force produced by a cylinder:

force produced by cylinder = air pressure x surface area of piston

force = pressure x area

If air going into a cylinder is at a pressure of 0.5 N/mm² and the piston is 50 mm in diameter, what force does the cylinder apply?

First calculate the surface area of the piston:

area of piston = πr²
= 3.14 x 25 x 25
= 1962.5 mm²

force = 0.5 N/mm² x 1962.5 mm²
= 981.25 N (Newtons)

Automatic Circuits

Many automatic circuits require a reciprocating movement (see page 62). By using air operated valves it is easy to build such a circuit. The speed or delay of any cylinder you use can be adjusted by adding flow regulators and reservoirs.

When building complicated systems, get the basic circuit working and test its operation first. Then add any speed control and delays later.

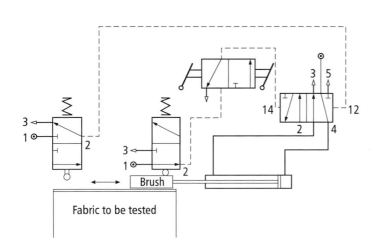
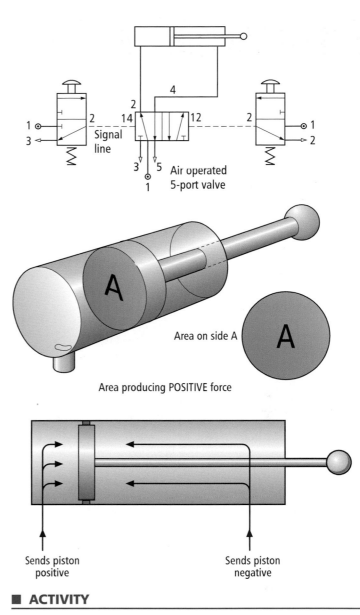

Area on side A

Area producing POSITIVE force

Sends piston positive

Sends piston negative

Air operated 5-port valve

Signal line

Brush

Fabric to be tested

■ ACTIVITY

Redesign the fabric tester circuit shown on the left, so the speed of the cylinder movement can be adjusted. Add a counter so the number of strokes can be recorded.

IN YOUR PROJECT

Pneumatics would provide a way of achieving the movement in your testing machine. Does the speed of the cylinder need to be adjusted? Look at how someone walks. You could attach a shoe or wearing surface to the cylinder. How would you then modify the cylinder?

KEY POINTS

- The speed of a cylinder can be controlled by a flow regulator.
- A cylinder's operation can be delayed by using a reservoir.
- Automatic control can be achieved by using air operated valves.
- It is possible to calculate the force applied by a cylinder.

HI-TECH HOTELS

pneumatics

61

A Moving Experience

There are four different types of motion or movement – linear, rotary, oscillating and reciprocating. One of the tasks often facing a systems designer is to convert one form of motion into another.

Spot the Motion

Linear motion is movement in a straight line.

Rotary motion is movement that just keeps on going around and around.

Oscillating motion is movement for swingers.

Reciprocating motion is from me to you, you to me.

If a system is operating a motor then the output motion will be rotary. This motion can, for example, be used to raise and lower a car park barrier by using a mechanism that will convert the rotary motion of the motor into a more useful form.

Introducing Mechanisms

Mechanisms are designed and used with one purpose in mind – to allow the user to carry out a task more easily. In our lives we are surrounded by different mechanisms that are used for a wide variety of purposes.

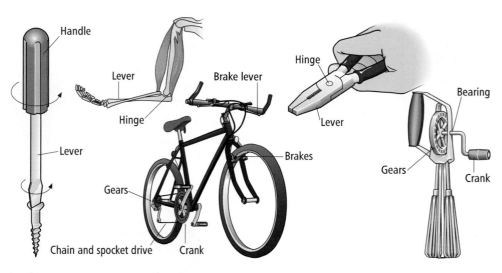

Mechanical systems

If we use a block diagram to describe the operation of a mechanical system it helps us to understand how the system can change the input motion into a different output. Consider the simple example of operating a typical front door lock.

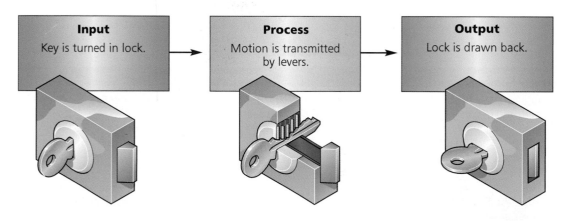

Input	Process	Output
Key is turned in lock.	Motion is transmitted by levers.	Lock is drawn back.

■ ACTIVITY

Describe the operation of each of the mechanical systems on the facing page using a block diagram. You may find it necessary to research into mechanisms further to be able to describe the process of each system fully.

IN YOUR PROJECT

► Explain the advantages and disadvantages of using each type of movement to provide the method of opening the door.
► State clearly which you think will be the most suitable.

KEY POINTS

● There are four types of movement – linear, rotary, oscillating and reciprocating.
● All things move by one of these four types of motion.
● A mechanism is used to change an input motion into a different output motion.

System Modelling (1)

When designing and making systems it is not always possible to construct fully operational working prototypes. However, designers still need to be able to demonstrate their ideas to others. The answer is to use system modelling components, some of which are described below.

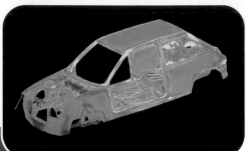

Industrial Modelling

The costs of developing a new car are extremely high. It takes many years to travel from the designer's first thoughts to a finished and working product coming off the production line. During this journey many ideas will have been suggested and accepted or rejected. For the designers to make the right decisions they need to use modelling systems. The most commonly used are mathematical modelling, physical modelling, graphic modelling and computer modelling.

Modelling system	Possible use	Example
Mathematical	Application of formulae to calculate the performance of a system, in part or in whole.	The calculation of the force acting on a pneumatic piston: force = pressure x area $= 0.255\,N/mm^2 \times 1962.5\,mm^2$ $= 500\,N$
Physical	The solid modelling of a design in order to see what the finished product would look like. Physical modelling materials also now include pre-formed construction kits.	The making of full-scale clay models of cars in order to see what the design would look like.
Graphical	This is the designer's way of communicating an idea to others. It might take the form of initial sketches or be formal drawings.	The use of design sketches to demonstrate first ideas, before going on to construct a physical model.
Computer	Designer's ideas can be modelled quickly and easily to see how they might operate in a variety of conditions.	Software programs that allow architects to see what shadows on the surrounding areas will be caused as a result of a proposed building development.

Modelling a System in School

There are many specialist system modelling kits available to the designer. Those suggested here highlight some of the more common examples readily available in schools. In most cases a modelled system will be a combination of fully working prototype sub-systems together with kit and self-made models and components.

Electronic sub-systems

Many schools are now able to produce good quality printed circuit boards, often with the supporting computer software to allow students to design or adapt their own circuits. The need to model an electronic circuit often needs to be met even if the facilities for prototype development exist. Component selections need to be tested to make sure that their mathematically modelled results are correct.

Some computer modelling software programs allow you to select existing circuits for modification or to design your own. These types of programs allow quick and easy changes in component values to monitor the effect on the circuit output.

Having modelled and refined a circuit on screen, you can model real components using circuit construction kits or prototype boards ('breadboards').

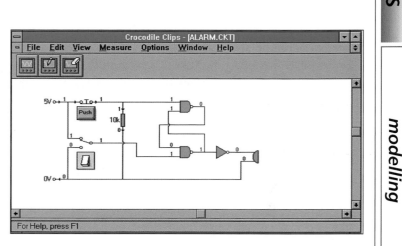

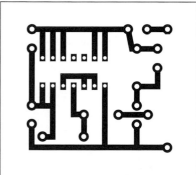

Pneumatic sub-systems

Designers will often model a system before full-scale construction because of cost constraints. It is better to find out if an idea isn't going to work as intended before you make it.

Pneumatic components are expensive, so in school any pneumatic solution to a design problem is likely to be modelled, as shown on the left.

ICT ➜

Consider the advantages of using software such as Crocodile Clips.

HI-TECH HOTELS

modelling

System Modelling (2)

Microprocessor sub-systems

'Computer control systems' usually refer to the box, monitor and keyboard systems that sit on many desks in offices and the home. As miniaturisation and technology develop, so more and more devices rely on microprocessor control. Microprocessors are the programmable components of computers.

Control boxes are readily available, allowing you the opportunity to develop simple computer controlled systems.

A PIC (Peripheral Interface Controller – see pages 35 and 94) can be programmed to control a sequence of operations and react to input sensors.

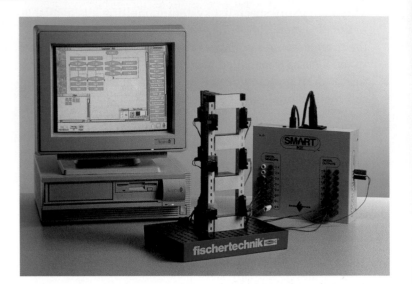

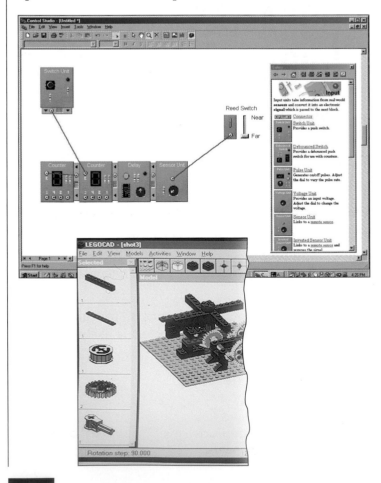

Mechanical sub-systems

The construction of fully working prototype mechanical systems is probably the most difficult of all systems to model. In school workshops it is often difficult to obtain and use the appropriate materials and to make sure that the components meet the necessary tolerances in order that they can operate efficiently.

As a result, it is common practice to manufacture mechanical sub-systems using commercially produced components. The nature of the system will determine whether or not the associated mechanisms are referred to as being 'modelled' or 'prototyped'.

Consider a design for a small robot arm. If a designer decides to use gear systems to provide the required movement, then commercially available plastic gear wheels might well be used. However, if the designer was developing ideas for a small lift system it is likely that plastic gears would be unsuitable for the actual device, but could be used in the development of a proposed gearbox.

Pulley systems often fall into the same category. String might be used to act as a lifting cable for a model, but would we really trust a lift system that used string, especially if we were a passenger in that lift?

ICT ➡️

You could use a digital camera to make a record of your models.

Structural sub-systems

Structural sub-systems can be modelled either by using commercially available kits or by making use of materials available in the school workshop.

Commercially available components that mimic real-life structural components have been available for over fifty years – from Meccano (which allowed structures to be constructed out of metal using nuts and bolts) to their modern day equivalents (e.g. LegoTechnic and FischerTechnic).

All such systems offer the designer a degree of flexibility, but they can be limited in that the components are designed to meet the needs of many. As a result they may not always meet the needs of the specific task.

It may be necessary to make a customised part for your structural sub-system using materials available in the school workshop. The task of having to make your own components can be helpful in understanding how material will react when it is used. For example, if a wooden beam is needed, you can compare the performance of a modelled beam using a variety of possible solutions. You might compare:

▷ beam sections
▷ the effect of grain direction in solid timber
▷ the possibilities of using manufactured materials such as plywood.

IN YOUR PROJECT

Having completed your design work consider which method of modelling will be the most suitable to demonstrate your solution. Will you need to use some commercial kits, can you make the components yourself, are electronics to be involved?

■ ACTIVITY

Consider the following mechanical devices. For each, suggest what material you would use to make the gear system. Support your suggestions with reasons.

▶ cable of a hotel lift
▶ sprocket system on a mountain bike
▶ gear system in a Formula One car
▶ gear system in a remote controlled car
▶ gear system in an electric food whisk.

KEY POINTS

● Models allow a designer to check an idea before manufacturing begins.
● They also allow a designer to communicate ideas to others.
● There are four types of models appropriate to Design & Technology: physical, mathematical, graphical and computer models.
● Models can be constructed using commercial kits or self-made components.

HI-TECH HOTELS

modelling

Safety First / Making It

Issues of health and safety cannot be under-estimated and must always be anticipated and planned for. It is the system designer's duty to make sure their solutions will operate safely and reliably. System operators also need to be aware of dangers and safety procedures when operating a device.

Think Safety

In this unit you have been studying a variety of technologies that involve movement and moving parts. All machines are potentially dangerous. The very fact that they have moving parts means potential injury is a factor that must be taken into account in their design and operation.

You have also been learning about pneumatics and hydraulics as a means of providing power to a system. Both have the potential to cause injury, and the need for safety when using them cannot be overstated.

General Safety

There are general safety rules that should be obeyed regardless of the equipment being used. We often refer to these as common sense rules.

Most accidents in the school workshop occur when individuals do not think and work in a safe and sensible manner.

▼ *Never run* in the workshop.

▼ *Dress sensibly* for the workplace. If you are using hazardous materials or equipment protect yourself in the correct way.

▼ *Wear safety glasses* or goggles when appropriate.

▼ Do not use *any* equipment unless you have permission.

▼ If in any doubt about safety at all, then *ask your teacher*, don't take chances!

Pneumatic and hydraulic safety

In a controlled situation pneumatic and hydraulic devices and systems are safe. If they are used incorrectly they can present a serious hazard to the user and those around.

Below are some simple safety guidelines for you to follow.

▼ *Never point a live airline at yourself or anyone else.* Compressed air is very powerful and can cause severe damage to the skin.

▼ *Protect your eyes with goggles or safety glasses* when doing practical work. A compressed air jet pointed at the eyes can cause very serious injury. Loose particles from the work area can also be blown into the face by a cylinder's exhaust.

▼ *Never try to stop components in motion with your hands.* Compressed air can store a great deal of energy and pneumatic components can have very powerful movements.

▼ *Make sure that all pipework is secure.* Loose pipes will move violently if air is passed through them.

▼ *Always consult your teacher* to check that your circuit is safe.

Safety with mechanisms

As with pneumatics, the use of mechanisms and mechanical systems in a controlled situation should not lead to the safety of the user and others being at risk. below are some simple guidelines for you to follow.

▼ When using machinery make sure that all working parts are properly guarded.

▼ On/Off switches must be clearly seen and within easy reach of the user.

■ ACTIVITIES

1. Look around your Design & Technology workshops and check out some of the machines. What safety features can you find on them? Make a table of your results comparing the different machines and their uses.

2. Safety signs in public places have to be designed to meet British and International Standards. In all of these, colour plays an important part. Find out what types of signs make use of red, green, yellow or blue as background colours.

3. Written safety instructions assume that the person looking at the message can read and understand it. Design a set of simple safety warning graphics that could be used without written instructions.

KEY POINTS

- When working as a designer you must ensure that all devices are designed with the safety of the user in mind.
- Safety in the workshop cannot be taken for granted. You must always think and work with safety in mind.
- Never put the safety of others at risk because of a thoughtless action.
- Remember, if in any doubt at all, ask. Never take chances!

IN YOUR PROJECT

What safety instructions and procedures will be needed for your system?

Making It!

Once you have completed all the design work, you need to present your solution as a fully working model for testing and evaluation.

Sub-systems

Consider what will be the most appropriate method of modelling for each of the sub-systems you have designed. What commercial kits can you use, and what components will you need to make yourself?

Planning the making

You will need to plan your workshop sessions very carefully. Develop flow diagrams which show the stages involved. Before you start, can you anticipate what might cause problems? What might you be able to do about them?

Testing

How will you go about testing your design?

In a project such as this reliability will be important, so how will you prove the reliability of your solution?

Remember, it will be important to test that your door operated as expected, as well as trying to test for unexpected conditions.

Final Evaluation

From your test results, what conclusions can you come to about the suitability of your design? Does your solution meet the requirements of the original specification?

Moving on

What further developments to your solution would need to be made or considered before work could begin on manufacturing the 'real' door?

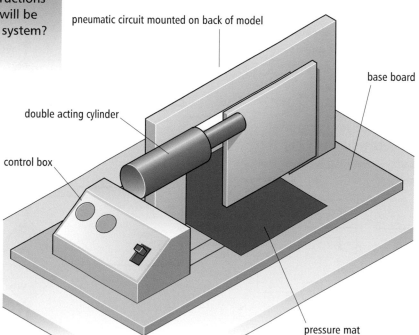

pneumatic circuit mounted on back of model

base board

double acting cylinder

control box

pressure mat

Project Two: Starting Point

Thinking Logically
(page 72)

Ventilation and air conditioning systems are becoming more common in our everyday lives. Maintaining a comfortable environment is important to ensure productivity in the workplace.

Can you design and make a system which will maintain a comfortable temperature in a kitchen?

Designing, Making
and Evaluating
(page 86)

If You Can't Stand the Heat

Read through the task below. Before you start to develop and finalise your suggestions for a design solution, however, you will need to work through the sections which make up this project.

Firstly you will need to develop your skills in logical thinking, and learn more about different sources of energy. Next you will look in more detail at process components (transistors and thyristors) and sensors. Finally you need to study motors and relay circuits.

The Task

Cooking up a dream

Future Hotels is well aware of the need to provide staff with facilities as good as those afforded to their guests. They want to make sure that the kitchen facilities in the Millennium Hotel meet the expectations of their chef and his team.

High temperatures in the kitchen can often lead to stress and mistakes being made. Future Hotels would like you to suggest how they might automatically maintain a good working environment in the kitchen.

System

Input	Process	Output
Temperature sensor	Switching circuit	Motor drives fan

Component

Input	Process	Output
Thermistor	Transistor	Relay switches motor

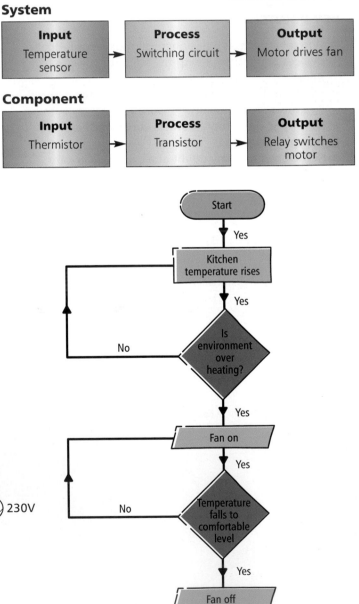

A flow chart to demonstrate the circuit problem.

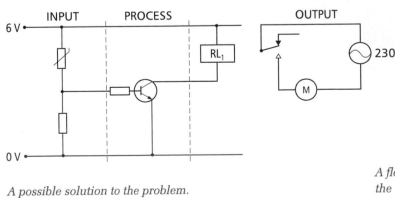

A possible solution to the problem.

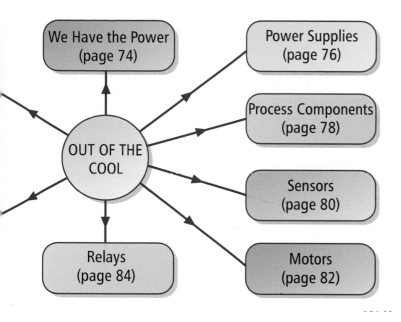

Starting Points

Before you start to develop your ideas you will need to consider the following.

Investigation

Find out what temperature levels are acceptable for hotel kitchens. Will your system need to comply with any Health and Safety legislation?

System control

What control will the users of your system have? Will they be able to override the input to stop the fan coming on, or to keep the fan on even when the temperature has fallen?

How will the system be switched on and off, or is it expected that it will always be running? Does this mean that there needs to be an 'on' indicator?

System process

What type of system process component will you need? Your choice may well be made once you have decided on the output.

What type of timing circuit would be suitable?

System output

You will need to decide whether or not you want your circuit to operate the fan directly. To be powerful enough you may need one that is mains operated. This will require switching by means of a relay.

System modelling

Before constructing your final circuit, try out different ideas by modelling first. Use a software program or a prototype system.

System casing

Given that your system is to operate in a kitchen environment, the casing will need to be suitable and able to withstand the conditions. Can the design of the casing indicate the function of the circuit? What methods are available to you for making your prototype? Would these need to be reconsidered if the system was to be mass-produced?

Having good ideas

In designing your solution you will need to think of ways in which the simple solution outlined on the facing page can be improved.

If the fan goes off as soon as the temperature falls to a comfortable level, does this mean that it will come on again immediately the temperature rises by a fraction of a degree? How can you make sure that the fan will reduce the temperature level well below the triggering point?

It might be possible to build some sort of time delay into the process section of the circuit. What do you think?

Off You Go!

As you work on the project you will find that there are many other issues to be resolved and decisions to be made. These need to be recorded and should form a part of your project folder.

SAFETY FIRST!

At no time should you use mains voltage for any part of a *Systems and Control Technology* project.

Thinking Logically

Flow charts are a form of modelling which describe the operation of a system in simple terms.

Truth tables are used to show the input and output of logic gates.

ICT

You could use CAD software such as Crocodile Clips to help learn how logic works.

Flow Charts

As well as using flow diagrams to describe how a system operates designers can also use flow charts to help them understand what factors will need to be taken into account when designing a system. They are a form of graphic modelling. When drawing a flow diagram it is important to use the accepted conventions.

Stage in system	Description	Symbol
Start	Used to indicate the beginning of the system's operation.	
End	Used to indicate the end, or termination, of the system's operation.	
Input/output	Used to show where an input to a system is received, or where there will be an output from the system, e.g. a motor is turned on.	
Process	The process box is used to show what the system is doing at that point, e.g. a temperature level is monitored.	
Decision	Used to indicate that at this point a decision has to be made by the system's control before its operation can continue.	

Flow Chart Symbols

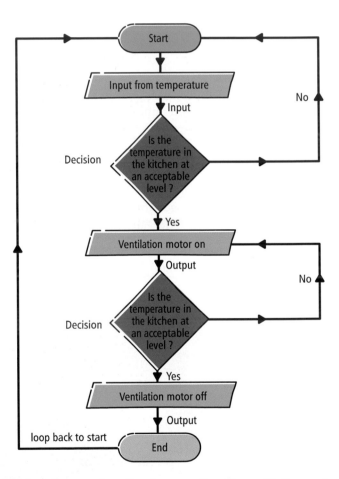

This flow diagram describes the operation of a ventilation system in simple terms. As the designer develops and refines the idea, the flow diagram will become increasingly complex.

The Truth about Logic

You will often need to describe the operation of the logic gate in a shorter form than drawing out the complete system. This is done by drawing out the truth table for the logic gate.

When using a truth table you refer to the inputs and outputs by their logic values.

Electrical equivalent	Logic value
Low or off	0
High or on	1

OR gate

Y	Z	Q
0	0	0
0	1	1
1	0	1
1	1	1

AND gate

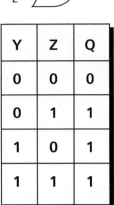

Y	Z	Q
0	0	0
0	1	0
1	0	0
1	1	1

NAND gate

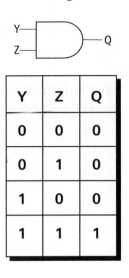

Y	Z	Q
0	0	1
0	1	1
1	0	1
1	1	0

Key: Y and Z = inputs, Q = output

Each logic gate has its own truth table. In the examples above the truth tables show only two inputs. This will not always be the case. A logic gate can have any number of inputs, but the truth table description will still have to meet the same rules.

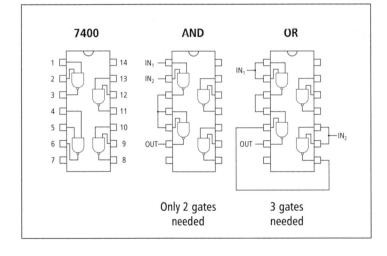

7400   AND   OR

Only 2 gates needed   3 gates needed

■ ACTIVITIES

A central heating system consists of a boiler, radiators, a hot water tank, a room thermostat and a timer that turns the system on and off twice a day.

► Write down the system as a logic description; refer to pages 54 and 55 for guidance.
► Draw out the system as a logic diagram; include the truth table for each gate used.
► Convert your logic table to show how one, or more, 7400 chips could be connected to perform the same function.

Combining Logic Gates

It is not always practical to use a variety of different logic gates in a circuit. The NAND gate, however, can help us to overcome this problem. The 7400 chip is known as a two – input quad NAND gate, and by connecting the inputs and outputs of the chip we can make the NAND gate perform the function of an AND gate and an OR gate, as shown on the left.

Using the 7400 chip to perform AND and OR functions.

IN YOUR PROJECT

A ventilation system needs to exhaust the air in the kitchen when it gets too hot. How might this problem look when modelled by a flow diagram?

NAND gates can be very useful in coursework projects, but will not be tested in the written examination

KEY POINTS

● Designers often use a flow chart to understand better how a system will need to operate, before they begin the process of designing.
● A truth table shows the inputs and outputs of a logic gate.
● Logic 1 means high or on. Logic 0 means low or off.
● We can connect a number of NAND gates together to form the equivalent of other logic gates such as an AND gate and an OR gate.

We Have the Power

If someone or something has energy then he, she or it, has the power to do work. Energy cannot be created or destroyed. As energy is transferred from one part of a system to another it can be made to do useful work.

Systems and Energy

On page 30 we looked at the four different system groups:

▷ natural systems
▷ designed systems
▷ abstract systems
▷ human activity systems.

Each of these requires some type of energy input. The system process then transfers this energy to produce the desired system output. It is often useful to use a shorthand way of describing energy in different systems:

▷ **potential**, e.g. a spring when it is compressed
▷ **kinetic**, e.g. a twisted elastic band when it is released
▷ **chemical**, e.g. the energy from fuels such as natural gas and oil
▷ **thermal**, e.g. a resulting from the transformation of chemical and electrical energy
▷ **electrical**, e.g. the energy transferred by a battery
▷ **light**, e.g. a lightbulb is used to transform electrical energy into light energy
▷ **sound**, e.g. turning up the volume on your stereo increases the sound energy
▷ **nuclear**, e.g. the release of energy as a result of nuclear fission.

The energy itself is not changed. The 'forms' listed above describe the systems which store or transfer the energy.

Energy Sources

Energy is commonly described as coming from one of two sources. These are:

▷ finite sources
▷ non-finite sources.

Finite energy sources

A source of energy that cannot be replaced once it has been used is said to be **finite**. Fossil fuels, coal, oil, natural gas and nuclear are all finite sources. With the exception of nuclear, these energy sources have been available for many years. If your school is still using this book in the year 2030 it is likely that many of these sources will have already been used up!

Non-finite energy sources

These are often called **renewable** energy sources. The energy produced by them is continually being replaced and will never run out. Examples of non-finite energy sources are solar, wind, wave and water. Our use of these sources of energy is often limited by our ability to make them efficient and cost-effective.

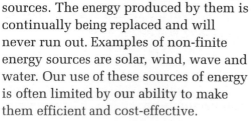

Energy Transformation

Remember, energy cannot be created or destroyed. Any designed system requires energy at its input and outputs energy in another form. In between the input and output the energy must therefore have been changed, or **transformed**.

This transformation might be along very simple lines.

Remote controlled car

Input energy	Conversion	Output energy
Chemical (battery)	Control circuits (electrical energy)	Kinetic (motor rotates)

Or it could be a more complex operation:

Electricity generation

Input energy	Transducer	Output energy
Chemical (coal/gas/oil)	Water heated to make steam (potential energy)	Electrical (electricity)
	Steam drives the turbine (kinetic energy)	
	Turbine rotates generator shaft (kinetic energy)	

Coping with Loss

One problem to overcome with all systems is how they cope with the loss of energy to the surroundings as a result of their operation. We are all used to electrical devices becoming warm as we use them, but what we don't always realise is that the transformation of this unwanted thermal energy is a waste of the input energy. The **efficiency** of a device or system is a measure of how much of the energy supplied to it can be transformed to the desired output. This is described by the ratio:

$$\frac{\text{energy out}}{\text{energy in}} \times 100\%$$

When selling a car the manufacturers make a great deal of the fuel consumption figures for their various models. If they claim that a car is capable of travelling 40 miles on a gallon of petrol they are in fact telling us how efficient their car is at transforming the input energy (in the form of petrol or diesel) into the output energy (kinetic energy as the car moves). With this in mind, many car manufacturers are looking into ways to reduce the weight of their cars and to make the engines more fuel-efficient.

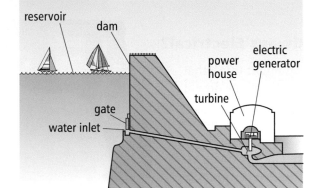

A hydroelectric power system transforms potential energy to electrical energy

■ ACTIVITY

1. One method of reducing the temperature level in the hotel kitchen might be to consider more energy efficient ways of preparing and cooking food. Carry out some market research into energy efficient devices for the kitchen. What makes them energy efficient?

2. Draw a simple flow diagram to demonstrate the energy transformation that take place in a hydroelectric generating station.

IN YOUR PROJECT

Identify the sources of unwanted heat that will be generated within the environment you are designing for.

KEY POINTS

- Energy cannot be created or destroyed, it can only be transferred or transformed.
- There are eight usable forms of energy.
- There are two sources of energy: finite sources that will eventually run out, and non-finite sources of energy.
- Most systems use one type of energy as an input and transform it into a different output energy.
- Nearly all systems suffer from energy loss to the surroundings. The less energy is lost, the more efficient is the system.

Power Supplies

Every system needs a power source for it to be able to operate. Designers need to make sure that their systems can be operated by an available and suitable power supply.

The main source of system power is usually electrical.

Always Electrical?

You might think that some systems do not need an electrical power supply to operate, for example a pneumatic system. However, even though a pneumatic system uses compressed air, this still has to be 'made'. This is achieved by using an electrical motor to pump air and compress it into a reservoir for use by the pneumatic circuit.

Electrical or electronic?

In simple terms, a device can be described as electrical if it normally requires mains electricity for its operation. Electronic systems are able to operate using the current and voltage provided by a battery.

Confusion often arises when an electrical item, such as a television set, is described as having electronic components. In fact, electronic circuits and microprocessors control the majority of new electrical items. So, how does an electrical device use mains electricity when it is controlled by electronic circuitry?

d.c. (direct current)
Direct current delivers a steady voltage which does not change its polarity.

a.c. (alternating current)
This is the name given to the electricity we use in our homes, provided for us by the National Grid system. We are able to use this electricity for devices fitted with the traditional 3-pin plug.

When it comes out of the sockets in our homes, the electricity is provided at a level of 230 V and has a frequency of 50 Hz. This means that the voltage level is 230 and that 50 times every second the voltage changes from a positive value to a negative one.

This is shown in the graph on the right.

Electrical devices can be controlled electronically by converting the mains a.c. into a lower voltage d.c. form. This is achieved in two steps outlined here.

Alternating current is converted into direct current by a process of **rectification**. In order to lower the input voltage we use a transformer to step down the voltage to a level that can safely power the electronic systems in the device. Full details of rectification and transformers are best found in physics or specialist electronic books.

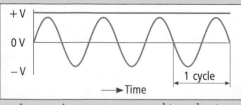

An a.c. sine wave compared to a d.c. trace.

a.c. versus d.c.
What are the advantages of using a.c. devices as opposed to d.c.?

▶ a.c. devices, such as radios and tape recorders, tend to be heavier than battery only, because of the weight of the transformer needed to turn the a.c. into d.c.
▶ Battery powered devices will always run out of energy so it is often necessary to make sure that spares are readily available.
▶ Motor driven devices, such as personal stereos, will drain a battery in a short space of time.
▶ Mains powered devices never run out of power, providing you pay your electricity bill!
▶ Battery powered devices are portable.

Perhaps the best solution is to choose a device that can be powered by both mains electricity and battery!

Direct current (d.c.)

Direct current is normally provided by one of a number of different battery types. The voltage levels of batteries range from 1.5 V to a maximum of 12 V. There are many types of batteries available, differing in construction and use. Most battery operated devices are designed to use an existing battery type, as shown below. In some specialist equipment, however, a custom-designed battery may be needed.

Battery type	Coding	Voltage	Amp hours	Typical applications
Zinc–carbon	3R12 PP9	4.5 V 9.0 V		Popular batteries for use as bench power packs and educational teaching applications.
Alkaline	AAA AA C D PP3	1.5 V 1.5 V 1.5 V 1.5 V 9.0 V	1175 mAh 2700 mAh 7750 mAh 18000 mAh 550 mAh	A range of long-life batteries often seen advertised. They are used to power many popular household items, such as torches, toys and radios.
Alkaline button cells	Many types	1.5 V		Sizes range from a diameter of 6.8 to 11.6 mm and a height of 2.1 to 5.4 mm. Used extensively in calculators, cameras, clocks, etc.
Lithium manganese coin cells	Various	3.0 V	From 50 to 260 mAh	Sizes range from 7.9 x 3.6 mm to 11.6 x 5.4 mm. Mainly used in miniature electronic equipment, such as calculators and watches.
Rechargeable batteries	AAA AA C D PP3	1.25 V 1.25 V 1.25 V 1.25 V 8.3 V	1200 mAh 1200 mAh 550 mAh 180 mAh 150 mAh	A typical rechargeable battery will usually have a minimum life of about 700 charge/discharge cycles. Uses are the same as those for alkaline batteries.
Nickel cadmium	Various	3.6 V 6.0 V 9.6 V	From 280 to 700 mAh 1800 mAh 1400 mAh	Available is a package design and style to suit a wide range of cordless telephones. Available to suit/fit a range of camcorder models.
Rechargeable lead acid	Various	6.0 V 12 V	1200–12 000 mAh 1200–24 000 mAh	These are maintenance-free rechargeable batteries. Uses include memory back-up systems, test equipment, fire and security alarms.

Information from Rapid Electronics Catalogue

Amp hours

A typical 3 V d.c. motor draws a current of 0.41 amps when operating at maximum efficiency. The **amp hour rating** (Ah) of a battery means that it will be able to deliver the stated current to a component for one hour. If the component needed only half that current, then the battery would be able to keep it working for two hours. Using this information, together with the rating of the battery, you can calculate how long the motor will run.

Battery type	Voltage	Amp hour rating	Motor current	Operating time calculation
PP3 alkaline	9.0 V	550 mAh	0.41A	$T = \dfrac{550\text{mAh}}{420\text{ mA}} = 1\text{h }18\text{min}$
Nickel cadmium	6.0 V	1800 mAh	0.41A	$T = \dfrac{180A}{420\text{ mA}} = 4\text{h }18\text{min}$

WWW. ➡

To find out more about batteries, go to:
www.duracell.com

IN YOUR PROJECT

▶ Find out what are the current ratings of the components in your circuit.
▶ What type and value of battery would allow your circuit to operate for the longest period of time without replacement?

KEY POINTS

● Most designed systems use an electrical current as their source of power.
● Electrical power can be supplied either from the domestic mains grid or from a battery.
● Mains electricity is supplied at 240 V AC.
● Batteries provide DC current in a range of voltage values from 1.5 to 12 V.
● There is a wide range of different battery types available .

Process Components

We have already considered a number of input devices. These need to be linked up with appropriate process components to make sure the system operates effectively.

Process Components Explained

In electrical and electronic systems the main purpose of the process component is to control the flow of current in the circuit. In schools there are commonly two components used to perform this function. They are:

▷ transistors
▷ thyristors.

When deciding which component to select, you need to be aware of the capabilities and advantages of each.

Data sheet: transistors	
Operation	This component acts as an amplifier. A small input current will activate the transistor causing a larger current to be sent to the circuit output. In this sense it is acting as a switch, although its operation is electronic rather than physical.
Popular transistors	There are two types of bipolar transistor that are popular for use in school projects:
▶ BC108	A small and sensitive transistor usually used when the output component is able to operate with only a small current.
▶ BFY51	A larger transistor that is not as sensitive as the BC108, used where the output component requires a larger operating component.
Field effect transistors (FETs)	When an even larger current has to be switched, a field effect transistor might prove to be more useful. One of the most commonly available is the IRF 530.

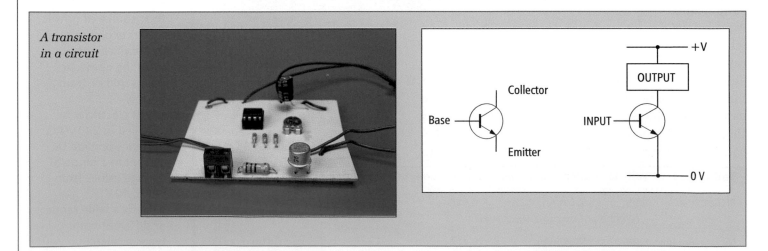

A transistor in a circuit

Choosing your transistor

The choice of transistor will depend on what output component your circuit requires. The following list should serve as a guide.

Use a BC108 if your output is:	Use a BFY51 if your output is:	Use a FET if your output is:
LED	LED	LED
Buzzer	Buzzer	Buzzer
Piezo sounder	Piezo sounder	Piezo sounder
	Bulb	Bulb
		Motor
		Solenoid
		Solenoid valve

Transistor combinations

The purpose of using a transistor is to create a circuit that can react to a small input signal. By combining a BC108 and a BFY51 we can increase the sensitivity of the circuit. The arrangement is known as a **Darlington pair**.

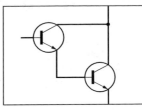

A Darlington pair

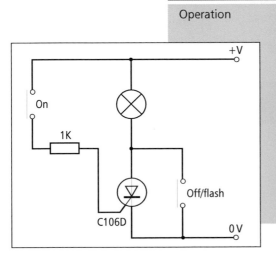

A thyristor in a circuit

Thyristor circuit outputs

A thyristor can be used to operate the same list of outputs that can be used with a FET.

Data sheet: thyristor	
Operation	A thyristor will also act as an amplifier but it will do more than simply switch on the output. When operated, it will also latch the output. This means that once the thyristor has been activated it will keep the output on, even if the input signal is removed or stops.
	A thyristor will operate as a simple latching component. The output can be switched on by momentary input pulse. It can be switched off by using another switch. This can be very useful in coursework projects, but will not be tested in the written examination.

IN YOUR PROJECT

Using a prototype board, experiment with the use of transistors and thyristors in circuits. Try using them with different input and output components. Is a Darlington pair arrangement really more sensitive? Record your findings for future use.

KEY POINTS

- There are two types of commonly used process components, transistors and thyristors.
- The choice of which component to use often depends on the type of output component that is needed.
- You can increase the sensitivity of a transistor circuit by combining two together to form a Darlington pair.
- Using a thyristor will allow you to latch the output as soon as it comes on.

More on Sensors: Finding the Right Level

The most commonly used sensors to detect a change in environmental conditions are the light dependent resistor and the temperature dependent resistor. They can also be used to measure levels of light and temperature to make sure that a circuit operates at the desired input levels.

The Light Dependent Resistor (LDR)

The most commonly used type of LDR is the ORP12. The resistance levels of this component rise when in darkness and fall when exposed to light. Exact resistance values vary from component to component, ranging from about 100 Kohms in darkness, to only a few hundred in light.

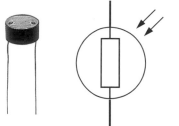

Measuring Light Levels

If we want to construct a circuit that will operate when a pre-determined lighting level has been reached, you need to be able to measure and quantify the level of light.

The problem
To design and make a circuit that will indicate when a bicycle rider should turn on the lamps.

The solution: 1
The designer needs to use a transistor switching circuit that will have an LED output. The circuit input will make use of an LDR in series with a resistor. This arrangement is known as a **potential divider**. The voltage across the base of the transistor determines if it is on or off. The proposed solution is shown below.

Input	Process	Output
Dark sensor	Transistor	LED

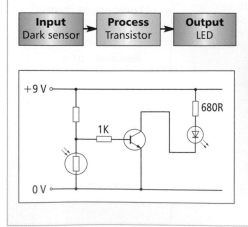

In order to construct the dark sensor we need to know what is the level of resistance in the LDR when the light falls to the level at which a rider needs to turn on the lamps. This value is needed so that the designer can select a value of resistor to go in the opposite half of the potential divided to the LDR.

Measuring the resistance values of an LDR is really quite simple. By connecting the LDR to a multimeter that has been set to read resistance, we can see the values change as the level of light falling on the LDR is varied. The following values were achieved by this method. At best these figures are only a guide.

Situation	Multimeter reading of resistance (typical levels)
Bright sunshine	270 ohm
Dusk	2 Mohm
A normally lit room	1 Kohm

Making decisions
Using this table of resistance readings, we can decide what value of resistor should go into the potential divided with the LDR. By using a variable resistor or pre-set, we can make final adjustments once the circuit has been constructed and we are able to test it in actual working conditions.

The solution: 2
The completed solution to the problem now looks like this.

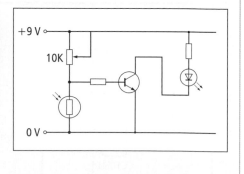

Too Hot to Handle?

When we want to have a nice long soak in a hot bath we might dip a hand in the water to tell us when the temperature is 'just about right'. In the majority of electronically controlled systems, however, 'just about right' is not good enough – we need a much more accurate detection system.

Accurate temperature detection systems can easily be constructed using a **temperature dependent resistor**, more commonly known as a thermistor.

The thermistor

A thermistor operates very much in the same way as an LDR. The most common types are those whose resistance fall when the temperature rises. There is a range of thermistors from which to choose. Details of these are given in the table below.

Thermistor type	Resistance at 25°C	Resistance at 100°C
Miniature disc 60–0300	300 ohm	30 ohm
Miniature disc 61–0320	100 Kohm	4 Kohm

Data from Rapid Electronics Catalogue

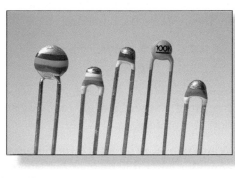

■ ACTIVITY

As with the LDR, we can measure the level of resistance of a thermistor by using a multimeter. Copy and complete the table below.

Situation	Temperature	Multimeter reading of resistance using a thermistor
Cup of coffee		
A warm room		
A frosty morning		

Using the results achieved above from your investigation, draw a circuit diagram that could be used in a device which is required to give a visual warning if the temperature outside the house has fallen below 0°C.

IN YOUR PROJECT

▶ Select a thermistor that is appropriate for the range of temperatures to which your device needs to be sensitive.

▶ Make sure that your selection of components is supported by evidence from the results of tests.

KEY POINTS

● The resistance level of an LDR varies depending on the amount of light shining on it.
● A thermistor's resistance level varies as a result of temperature.
● We can use both LDRs and thermistors to measure heat and light levels.
● Both an LDR and a thermistor are used in a potential divider, to act as an input to a circuit.

Motors Make the World Go Around

Imagine, if you can, life without motors. How many everyday items would cease to exist? We live in a world that needs motors to keep it running, but what is a motor and how does it work?

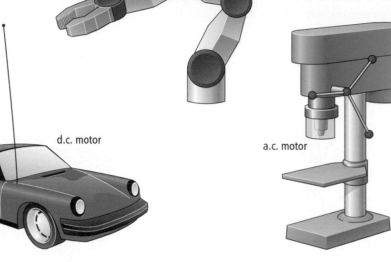

Stepper motor

d.c. motor

a.c. motor

Types of Motors

Motors come in many different shapes, sizes and types. The three most common types used by the systems designer are:

▷ the stepper motor
▷ the d.c. motor
▷ the a.c. motor.

In school you will mainly be using stepper and d.c. motors.

When choosing a motor a designer will first need to consider what it is needed for. What **torque** will the motor need to have? The term torque refers to turning forces, and when describing a motor we use it to indicate the turning force that the motor can exert.

The designer also needs to consider its power source: d.c. or a.c.?

Size might also be important if the motor is to operate in a confined space.

Using d.c. Motors

When a d.c. motor is used as an output device in a circuit its operation is usually controlled by an electronic circuit, rather than being directly switched on and off by the user. Because of this it is normal practice to use the circuit to operate an interfacing device, such as a relay (see page 84), that will turn the motor on and off.

d.c. motor outputs

The motor in itself is only one part of a circuit's output; in fact all that happens when a motor is turned on is that the spindle will rotate. In order to make the motor do some useful work a mechanism needs to be attached, or other component that can make use of the rotary output of the motor.

A remote controlled vehicle can be made to move or steer by attaching a gear system. This allows the reduction of the high output speed of the motor to something more useful. The addition of a fan blade onto the spindle will create a flow of air.

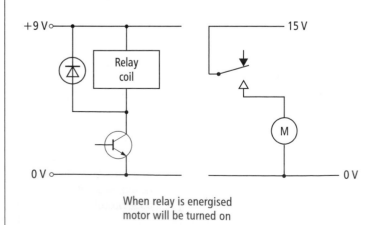

When relay is energised motor will be turned on

electronic components

Data sheet: motors

	Stepper motor	Miniature d.c. motor	a.c. motor
Power supply	6–12 V d.c.	Typically 1.5–3.0 V d.c.	Mains operated
In brief	A stepper motor rotates for part of a turn every time it receives a voltage pulse from its driver circuit. As a typical example the SM42 stepper motor will turn 7.5° every pulse, and will complete one revolution every 48 pulses.	Although we might be designing a system that requires a high torque motor, it is normal in school to model the solution using miniature d.c. motors. These are available in a variety of torque ratings and output speeds.	a.c. motors are available in a very wide range of sizes and specifications. The torque requirement of a lift will be much greater than that required of the motor that drives a washing machine.
Typical applications	Used where accurate positional control is needed, e.g. plotters, printers, etc.	Used in battery operated toys such as remote controlled vehicles.	Extensively used in industrial machines such as lathes and drills. High torque motors are also used to drive lift systems in shops and hotels.

Technical data of examples

Stepper motor			Miniature d.c. motor	
Step angle	7.5°		**1.5 V**	**3.0 V**
Holding torque	50 g/cm (50 mNm)	No load speed	8700	14000
Pull-in rate max.	320 steps/seconds	current (A)	0.32	0.38
Phase (coil) resistance	50 ohms/phase	At maximum efficency:		
		speed	5800	9400
Current per coil	258 mA	current (A)	0.76	1.1
Rotor inertia	13 g/cm	torque (g/cm)	5.3	8.6
Nominal voltage	12 V	efficiency (%)	32	30.5
		Stall torque (g/cm)	16.0	26
		Weight	17 g	17 g

■ ACTIVITY

Make a table of the different motors that are to be found in your school workshop. Find out the following information.

▶ Power: d.c. or a.c.?
▶ Use: what device or machine is the motor used on?
▶ Technical specification: you may need to carry out some research to find this out.
▶ Spindle attachment: what does the motor drive?

Illustrate the system/device using a block diagram.

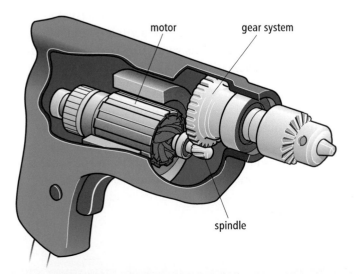

motor   gear system

spindle

IN YOUR PROJECT

▶ Construct a simple test rig that will enable you to experiment with different designs of fan blades to be driven by a motor.
▶ How can you demonstrate the efficiency of your design?

KEY POINTS

● A motor is an electrical device that provides rotary motion.
● There are three basic types of motor, a stepper motor, a d.c. motor and an a.c. motor.
● In school you will usually use stepper and d.c. motors.
● The torque of a motor refers to the turning force the motor can exert.
● When operating in a circuit it is advisable to give the motor its own power supply.
● In order for the motor to do useful work a mechanism or device needs to be attached to the motor's spindle.

Relays: Switching the Output

To operate an output device such as a motor or a solenoid we need to provide the device with its own power supply in order to protect the operating circuit. The simplest way to achieve this is to use a relay.

DIL reed relay

Miniature relay

Continental relay

The Relay

A relay is an electromagnetically operated switching device that is used to switch on and off output components that require a larger operating current than the control circuit can safely cope with. A relay consists of two distinct parts, the relay coil and the switching contacts.

Choosing your relay

The choice of relay will depend on a number of circumstances:

▷ What is the operating voltage of the relay coil?
▷ What current do the relay switch contacts need to carry?
▷ What type of switch arrangement is required?

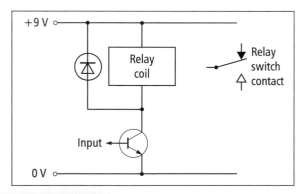

Switching the relay

In your GCSE course you will be learning how to use transistors and ICs, such as the 555 chip, as the process components in a circuit. We can use these to switch the relay coil on and off, and it is the coil that switches the relay contacts.

The relay coil operates as an electromagnet and we say that the relay is 'energised' when the coil is activated. The energisation of the relay causes the switch contacts of the relay to change over. It is this switching that controls the final output of the circuit, such as a motor.

Depending on the type of relay used, the switch contacts can turn on and off a wide range of output devices. This can include such things as motors and solenoids that require a different operating voltage than that of the input and control elements of the circuit. In some cases the relay switch contacts can even be used to control mains voltage.

IN YOUR PROJECT

To obtain a high grade for your coursework you will be need to combine a digital control system with a mechanical or pneumatic outcome. A relay or solenoid can help you achieve this.

Relay Switching

The switch contacts, unlike manually operated switches, are only available in two arrangements:

▷ single pole, single throw (SPST)
▷ double pole, double throw (DPDT).

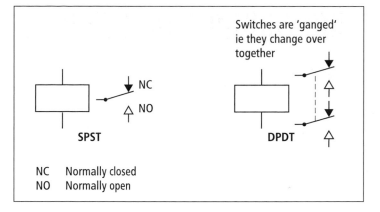

NC   Normally closed
NO   Normally open

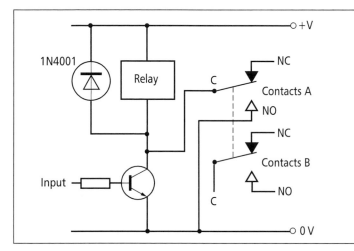

Relay latching

When using a DPDT relay we can use one set of switch contacts to keep the relay energised, even when the signal to the coil of the relay has been removed. This is particularly useful in alarm circuits where the output of the circuit can be kept on even though the original triggering of the circuit may have been removed. This is called **latching**. It is achieved by using one set of the relay switching contacts to bypass the connection of the relay coil with the negative rail in the circuit.

Electromagnets

Electromagnets are used in a variety of applications. Perhaps the most commonly seen are in scrap metal yards where they are used to move the scrap around the yard.

The advantage of the electromagnet is that the it is only energised when a current flows through the coil. The moment the current is stopped then the magnetism is lost.

■ ACTIVITY

Using an electronic catalogue, such as *Rapid Electronics*, produce a table that compares the following information of a range of different relays:

► relay name
► contact rating
► contact arrangement
► contact material
► mechanical life
► operate/release time
► coil consumption.

IN YOUR PROJECT

Explain why you need a relay that can be operated by an electronic input, but run a mains voltage motor.

KEY POINTS

● A relay is an electromagnetically operated switching device.
● A relay is used when the output device of a circuit needs a current beyond that which the process circuit can safely carry.
● When the relay coil is on, the relay is energised.
● Relay switching contacts are available in SPST and DPDT arrangements only.
● In a DPDT relay one set of switch contacts can be used to latch the relay coil in order to keep the output device on.

HI-TECH HOTELS

electronic components

85

Designing, Making and Evaluating

Choosing the right components for a system is critical if it is to work reliably and efficiently. When designing an integrated system the individual sub-systems can be designed first, and then linked to form a complete solution.

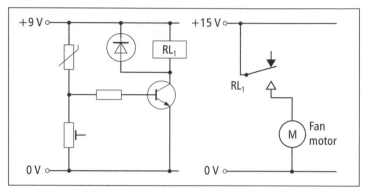

Component Selection

Designing a suitable circuit is only the starting point in achieving the aim of producing a successful solution to the problem. The components must be able to operate and perform their duties in the environment in which they are to be used.

Investigate available fans and decide which will cope with the task.

In a commercial situation the fan would be mains operated: you are aiming to produce a modelled solution.

Circuit Design

Once you have decided on the model of fan to be used you can select a suitable relay. There are some suggestions on pages 84 and 85 as to how an appropriate circuit might be developed. Given the information on the previous pages you should now be able to design a suitable solution to the problem. A suggested solution is shown above.

Input components
To choose the right thermistor for your circuit you need to know what temperature range it will be operating in.

What is considered to be a high temperature in a kitchen? Will it be the same as the temperature levels that you might want to maintain in a lounge, for example?

Information please!
Investigate the normal operating temperatures for kitchens; compare the temperatures in a domestic kitchen with a commercial kitchen such as the one in your school. From your results, can you suggest a suitable thermistor type?

Process components
A transistor switching circuit is suitable for this project, but which transistor should you choose? Remember, the choice of transistor needs to be made with the knowledge of the output device that the circuit is to have.

Output components
This circuit will need to have two output components: a relay that, in turn, operates the extraction fan. In choosing the type of relay to be used you need to know some details about the fan to be used in your circuit.

More information please!
When selecting a fan you will need to make a comparison of fan specifications that includes the following information:

► operating voltage
► input current
► airflow
► noise level.

Sub-system Identification

This project contains a number of sub-systems, each of which presents its own problems.

▷ The operating circuit:
 – the input (temperature sensor)
 – the process (monitoring and electronic switching circuit)
 – the output (the relay switching circuit operating the electric motor).

▷ Casings for the above:
 – the power supply and switches
 – allowing for access to replace batteries, etc., and for manual control of the device.

▷ A model to support the operating circuit and fan to demonstrate its use.

Testing Ideas

Make sure you have fully considered all the possible alternatives for each of the sub-systems. You may need to test out these trial solutions to your problem before you make final design decisions.

▷ Once you have decided on all of the sub-system designs, you will need to consider all interfaces.
▷ How will the control circuit fit into the casing?
▷ What controls will be needed, and how will these be fitted into the casing to ensure ease of use?
▷ How will the casing fit onto the model?

Don't let anything happen by chance. Try to anticipate all design decisions that will have to be made and plan to solve them.

You need to decide if a modelled solution will be sufficient to test and evaluate your final design, or if you need to make a fully working prototype.

Planning the Making

▷ Have you prepared a detailed component and materials list?
▷ Is everything you need readily available or will items need to be ordered?

Issues such as these can often cause delays during making.

ICT ➡
You could use a word processor to write up your testing and evaluation.

Final Testing

Set your circuit working.

▷ Is the input of the circuit sensitive enough to react to a change in temperature at the desired levels?
▷ Is your circuit efficient, reliable and able to operate in a range of different environments?

When carrying out your test, make sure your results are fully documented and analysed.

Final Evaluation

Think about your final solution. What conclusions can you draw from the testing?

▷ How well does it work? Are there areas you need to improve?

▷ What other applications could it be used for? Instead of sensing how hot the room is, could it be modified to sense how cold it is? Could the circuit be developed to draw air into a room as well as expelling it?

▷ Think about using a DPDT relay to reverse the current to the fan motor. Will the fan blade need to be changed?

▷ What changes and developments would need to be made if your final design was to be commercially produced? What manufacturing methods would you suggest to produce your design in much greater numbers?

▷ Have you produced a plan detailing the sequence of making for each sub-system, as well as ensuring successful completion of the whole project?

Examination Questions

Your teacher will tell you which of the following questions are appropriate to the focus areas in your course. You will need some A4 and plain A3 paper, basic drawing equipment, and colouring materials. You are reminded of the need for good English and clear presentation in your answers.

1. This question is about systems.
See pages 30-35. *(Total 6 marks)*

Draw a system block diagram for each of the following devices:

a) Braking system on a bicycle
(3 marks)

b) Door lock *(3 marks)*

2. This question is about sub-systems. See pages 36-45.
(Total 10 marks)

A burglar alarm system uses a number of sub-systems.

a) What is meant by a sub-system? *(1 mark)*

b) Name the sensor that is used to detect if a window is opened.
(1 mark)

c) Draw the system block diagram for a domestic burglar alarm system *(4 marks)*

d) If there is a power cut the alarm system will still need to operate. Name a suitable power supply for the alarm system.
(1 mark)

e) In recent years there has been a growth in the sales of burglar alarms. Give two reasons for this.
(2 marks)

f) Describe one problem with the growth in home alarm systems.
(1 mark)

3. This question is about sensing.
See pages 80-81. *(Total 13 marks)*

The circuit below is designed to switch on a fan when it becomes too hot.

a) Name and state the function of the components A, B and C.
(6 marks)

b) Redesign the circuit so it would switch on the fan when it becomes cold. *(2 marks)*

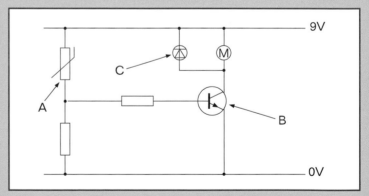

c) Before you would assemble the circuit by soldering the components, how could you ensure that it will operate as you want?
(2 marks)

d) Computer Aided Design (CAD) is now widely used in the design and manufacturing industry. Explain a specific example where CAD could be used in the design and manufacture of a PCB and the benefits it will bring. *(3 marks)*

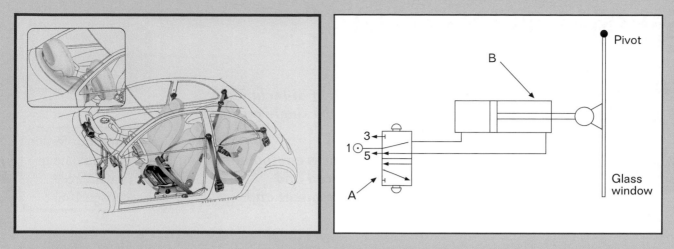

4. **This question is about logic.**
See pages 72-73. *(Total 14 marks)*

A car manufacture wants to add a built in safety feature so the car will not start unless all the doors are shut and the passengers are wearing a safety belt.

a) What logic gate would you use in this system? *(1 mark)*

b) What sensor could be used to sense if the door is closed? *(1 mark)*

c) Design another safety feature for a car. Describe how it would work. *(3 marks)*

d) Describe the advantages that modern CAD/CAM systems have brought to the automotive industry. *(5 marks)*

e) Before a car manufacturer would sell any new model it must carry out a series of tests. Describe two of these tests and the reasons for them. *(4 marks)*

5. **This question is about using pneumatics.** See pages 56-61. *(Total 20 marks)*

The photo below shows a pneumatically controlled greenhouse window.

a) Name and state the function of components A and B shown on the diagram at the top of the page. *(4 marks)*

b) When this was first used the window opened too quickly and was broken. How could the speed of movement be reduced? *(2 marks)*

c) The manufacturer wants to make an automatic pneumatic system, so the window opens and closes as the temperature rises and falls. Draw the electronic circuit diagram for this. *(4 marks)*

d) A force of 20 N is required to open the window. If the diameter of the cylinder is 40mm, what is the minimum air pressure required to open the window. *(4 marks)*

e) Design a pneumatic system that could open and close the door to the green house. *(4 marks)*

f) Give one advantage of using a pneumatic system in a greenhouse, when compared to an electro-mechanical system. *(1 mark)*

g) State a health and safety issue that must be considered when using pneumatics. *(1 mark)*

Project Three: Starting Point

The modern fitness gym is full of hi-technology equipment all designed with the single purpose of developing the user's body.

Can you design an item of fitness equipment which is suitable for experienced and inexperienced users, and by frequent and infrequent users at a hotel?

Investigating
Existing Structures
(page 92)

Assembling, Testing
and Evaluating
your Circuit
(page 106)

Designing and
Making an
Electronic Circuit
(page 104)

Getting into Shape

Fitness and exercise is a growth industry. Fitness equipment has come a long way since the days of simply lifting weights. Specialised fitness centres are opening up in every town and city, filled with the latest developments in equipment system design intended to stretch, pull and twist every muscle in the human body. The fitness industry has gone hi-tech.

Getting Fit

The human body is not as strong as we might like to think it is: we can only stretch and move in certain ways without causing ourselves injury. Most exercises are designed to develop the muscles in our body and the movements and exercises to do this have been known for many thousands of years.

Modern technology has helped designers to create machines that allow these exercises to take place more efficiently and safely, allowing the exerciser to make greater progress in their quest for the perfect body.

Preparing for the Task

Read through the task on the right. Before you start to develop and finalise your suggestions for a design solution, however, you will need to work through the sections which make up this project.

First you will need to make a detailed study of some existing examples of fitness equipment. Then you will need to learn more about structures, sensing, and electronic counters.

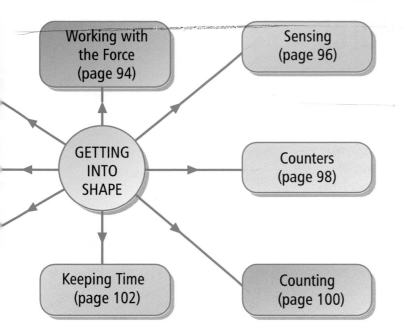

FUTURE HOTELS

Guests of Future Hotels have become accustomed to enjoying the benefits of our leisure facilities, and it is our intention to make sure that the Millennium does not disappoint them. We are therefore planning to install a state-of-the-art fitness centre as part of the Hotel's leisure complex. Safety is of prime Importance and we need to ensure that the equipment we have ordered will meet the demands of our guests.

In addition to this we want to enhance the equipment on offer by including monitoring equipment to enable our guests to plan their sessions in the fitness centre. Please respond to the following:

1. Provide an analysis of current fitness equipment with particular reference to:
   - material type and form used in their construction
   - an outline description of how the device operates from the point of view that it is a system consisting of a number of small mechanical sub-systems.

2. Suggest how guests using the equipment might monitor their performance by:
   - being able to electronically set a repetition target across the range of equipment we expect to install.
   - a simple timing device that has the flexibility to be used on a variety of machines within the fitness centre.

3. Could you please submit an outline design for a small piece of fitness equipment for use in the health centre at the Millennium Hotel. Model your proposed solution using either commercial construction kits, and/or self-made components.

Outline Specification

Fitness activities
The equipment will only be required to support a single fitness activity. The choice of the activity is left to you, although background information on the benefits to the user of the activity will be expected.

Dimensions
The overall dimensions of the device must not exceed 2 m in height, 1.5 m in width and 1.25 m depth.

Users
The device should allow for users from the following groups:
▶ adult males 18+
▶ adult females 18+

All decisions regarding dimensions should be clearly supported by anthropometric data and the final device should be ergonomically designed.

Materials and manufacture
The choice of materials is left to your discretion, although all choices made must be supported with detailed reasoning. The device can use either permanent or non-permanent fastenings. Clearly outline how your design would be manufactured and assembled.

Mechanisms
Clearly show how your system allows for the lifting of the weight. Indicate all of the components needed to complete your lifting mechanism.

ICT

You could use the bullet facility in a word processor effectively when writing your specification

Investigating Existing Structures

When designing the framework for a piece of fitness equipment it is important that the designer is fully aware of what the frame needs to achieve. Studying existing solutions can help provide useful information and ideas on important questions about the design.

Designing to Get Fit

In designing a piece of fitness equipment the designer will need to research and consider a wide range of factors:

The exercise
The designer needs to know what the exercise is in terms of the human movement that will need to take place, e.g. statistical information on the range of intended users in the form of anthropometric and ergonomic data.

Depending on the piece of equipment being considered, the designer will need to know the range of weights needed for a specific exercise, or the range of speeds that will satisfy the range of possible users on a jogging machine.

Stability
Will the device be free-standing or can it be fastened down to the floor? When the equipment is in use, will the movement of the user or parts of the device cause it to become unstable?

The user
What will be the position of the user when using the equipment? Will they be sitting down, lying on their back or front, or standing up?

What movements will the user be making when using the equipment? This will need to be taken into account when designing the frame of the device.

The operation of the equipment
Each piece of equipment is used in a different way. The designer needs to make sure that all of the necessary mechanisms and components can be securely held and contained within the frame.

The space available for the equipment will also have a bearing on the above points.

Safety
The safety of the user must always be an overriding concern in any designed system. In the case of fitness equipment, the user needs to be able to use the device in the confidence that the machine can safely support the activity.

Material and component choice
To begin with the designer will need to consider the different sub-systems of the device:
► the frame
► any mechanisms
► control features.

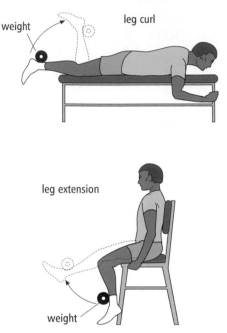

In each case the designer must be able to choose the right materials and components for the sub-system.

Cost
One piece of equipment will not be enough to equip an entire fitness room and so the cost of the equipment will be a major consideration.

Size
Fitness equipment often needs to be designed so that it takes up minimum space. This allows more equipment to be placed in a room, allowing more users at one time, which in turn means more money for the centre.

Versatility
Instead of having one machine for each exercise, it is often preferable to have a single piece of equipment that can allow users to perform a series of exercises simultaneously.

Investigation

The exercises

Most exercise machines are designed to allow one of two particular types of physical development:

▷ muscle toning
▷ aerobic training.

Muscle toning exercises are aimed at developing specific areas and require the user to be able to sit, stand or lie in a position where they can best carry out the exercise. Aerobic training includes the use of jogging and running machines, rowing machines and fitness bikes.

Statistical data

To ensure full use, all fitness equipment must be able to cope with a range of different user types, male and female, light and heavy, tall and short. Some method of adjustment will be needed to allow for this. You will need to know the capabilities of the users, in terms of the weight they can lift or the speed at which they can cycle.

■ ACTIVITIES

1. Find out all you can about the different types of muscle toning exercises. Make sure that you get information on the position of the user during the exercises and what movements they need to be able to carry out when exercising.

2. Collect any relevant anthropometric data that you feel may be of use when designing an item of fitness equipment. What ergonomic details also need to be considered in the design of such equipment?

3. Make a study of an existing item of fitness equipment. If you are not able to find a real example, refer to the photographs on this page. Answer the following questions:

Materials
▶ Name the main materials used.
▶ What properties do these materials possess that make them useful for this application?
▶ How have the materials been used and shaped?
▶ What other materials are used that do not form a working part of the structure?
▶ What sections are being used?
▶ What effect can the shape of the section have on a structure?
▶ Why do you think the chosen sections have been used?
▶ What finishes have been used and applied, and why?

Structural stability and rigidity
▶ What design features have been included to make the structure stable? How do they work?
▶ What other methods could have been used?
▶ What features have been included to make sure that the structure will remain rigid once it is in use?
▶ Could other methods have been used? If so, what are they?
▶ How has the designer decided to join the framework together?
▶ When would you use 'permanent' and 'non-permanent' fastenings in preference to the other?

IN YOUR PROJECT

A knowledge of structures can be very useful in coursework projects, but will not be tested in the written examination.

Working with the Force

A designer needs to be able to identify the forces that will be acting on a structure and make sure that its design can withstand these forces.

All structures that form part of a working system need to be able to accommodate two types of forces: **static forces** which are those created by the system itself, and **dynamic forces** which are created when the system is being used.

Identifying the Force

There are five forces that could act on a system, and in the case of fitness equipment all of them are represented. The forces are:

▷ compression   ▷ tension   ▷ bending.
▷ torsion   ▷ shear

■ ACTIVITY

Study the photograph on the left. Use the following descriptions to identify the types of force acting on the device.

A The user sits on the seat, which is supported by two uprights. What force must these resist in order for the seat to hold the user?

B The user pulls down the overhead bar and the lifting cable is connected to the centre of the bar. What type of force will this cause to act on the bar?

C When in use, the weight is raised by the attached cable. What force must the cable resist if the weight is to be raised?

D The overhead arm holds the top pulley in place, allowing the user to pull straight down on the overhead bar (B). What force does the overhead arm have to overcome in order to hold the pulley in the correct place?

E As the weight is raised, all pulleys in the system must be held securely in the correct positions by their fixings to the frame. What force must the pulley fixings overcome in order to achieve this?

Making the Effort: Lifting the Load

The need to conserve space with fitness equipment has become an increasing demand made on the designer. The response has been to develop systems that include some complex mechanisms to transfer the effort of the user to lift the selected load.

In designing such a mechanical system the designer needs to be aware of the movement that the user will make in order to carry out the exercise. For example, to perform a leg curl the exerciser needs to be lying down on the stomach, able to raise the legs from the knees upward to a vertical position.

In most examples of fitness equipment, pulley and cable lifting systems are used, because these allow the direction of force to be transferred without increasing the effort needed by the user. By clever use of the pulley mechanisms, the space needed for the device can be restricted.

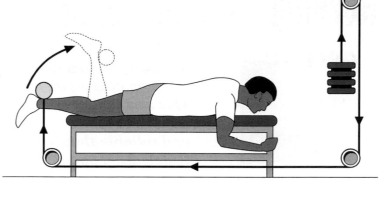

 Pulley points

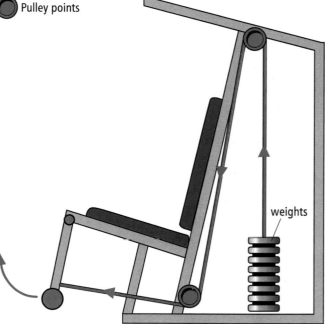

weights

■ ACTIVITIES

1. Answer the following questions for each of the examples drawn on the facing page.

▶ How does the material chosen help to overcome the effect of the force?
▶ Suggest an alternative material to the one chosen.
▶ What are the static forces acting in the system?
▶ What dynamic forces does the structure need to be able to withstand?

2. Many different devices make use of pulley lifting systems. Compare a number of different systems, commenting on the following:

▶ the purpose of the system
▶ the load to be raised
▶ the materials used for the pulleys
▶ the materials used for the pulley cable.

What different power sources are used to drive the lifting system?

IN YOUR PROJECT

▶ Decide for what fitness exercises your device is to cater. Draw line diagrams of the movements that will need to be made by the user.
▶ Decide where the weights will be. Draw line diagrams showing the possible paths of the cable in order for the movement of the user to lift and lower the weights.

KEY POINTS

● There are five forces that act on all structures. These are compression, tension, bending, torsion and shear.
● Structures must resist two types of forces, static and dynamic.
● Pulley mechanisms are often used to transfer the direction of an effort when lifting a load.

Sensing

It is possible to make a sensing circuit by using an operational amplifier or op-amp. This component can be used in a number of different ways. An op-amp is more sensitive and reliable than a transistor. The 741 is a widely used op-amp.

The 741 Op-amp

An op-amp has two inputs. One is an inverting input which is pin 2 and is marked as negative. The other is a non-inverting input which is pin 3 and is marked as positive. There is only one output. The 741 is packaged in an 8-pin DIL case, which looks like a 555 timer IC. You can find out more about the 555 on pages 96 and 97.

The 741 needs to be powered by two 9 volt batteries, which can supply the required −9V, +9V and 0V. This can be achieved by joining the two batteries together as shown.

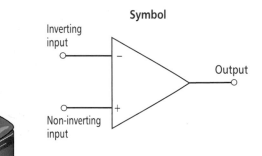

Symbol

Inverting input

Non-inverting input

Output

+9V

0V

−9V

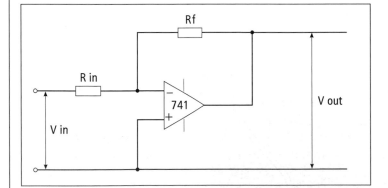

Rf

R in

741

V in

V out

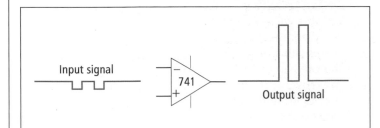

Input signal

741

Output signal

There are two useful ways of using the 741. One is as a voltage amplifier and the other is as a comparator.

The op-amp as an Inverting Voltage Amplifier

The op-amp will amplify an input voltage. If the output voltage is 20 times greater than the input voltage, then the gain of the amplifier is 20.

The gain is controlled by the values of the input resistor R_{in} and the feedback resistor R_f.

$$\text{gain} = \frac{V_{out}}{V_{in}} = \frac{-R_f}{R_{in}}$$

If R_f is 10 K and R_{in} is 1 K, then the gain would be 10.

The circuit is called an inverting amplifier because the output voltage is the opposite polarity to the input voltage. This is shown by a negative in front of R_f.

The Op-amp as a Comparator

It is possible for a 741 to compare two input voltages. This is known as a comparator.

If the voltage going into the inverting negative input is greater than the non-inverting positive input, then the output will be a negative voltage.

If the voltage going into the non-inverting positive input is greater than the inverting negative input, then the output will be positive (see table on right).

If the input voltages are equal, then the output will be zero.

Positive input voltage	Negative input voltage	Output
Large voltage	Small voltage	+
Small voltage	Large voltage	–
Equal voltage		0

Sensing Temperature

This type of circuit is sensitive and will sense small changes of input voltage. It can be used with a sensing component, such as a thermistor. As the temperature changes, the internal resistance changes in the thermistor, and this in turn will change the voltage going into pin 2 of the 741.

The value of the variable resistor VR can be adjusted to control the point at which the output signal changes. This type of circuit can be used to control an automatic heating system. As the temperature rises and falls, the voltage going into pin 2 will rise and fall above the pin 3 voltage, which is set by VR. This in turn will switch the output on and off.

741 op-amp

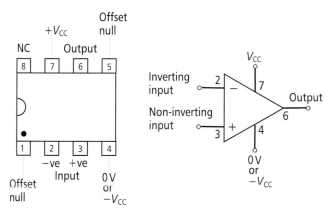

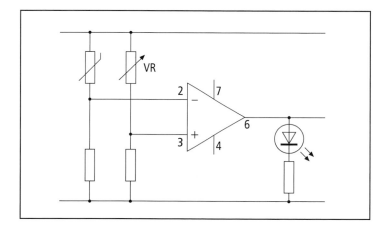

You could use CAD software such as Crocodile Clips to learn how an op-amp works.

■ **ACTIVITY**

Design a light sensing circuit using a 741 op-amp. Test its operation by making the circuit on prototype board.

IN YOUR PROJECT

▶ Why would it be useful to control or monitor accurately the temperature in your fitness centre?

▶ Design a circuit that could carry out the need you have identified.

KEY POINTS

● The 741 op-amp has two inputs, inverting and non-inverting.
● The 741 can amplify an input voltage.
● The 741 can compare two separate input voltages.
● The input voltages can be controlled by a sensing component, such as a thermistor.

Counters

In many systems it is necessary to keep a record of the number of operations carried out. This can be achieved by using a counter. Counters can be mechanical, electromagnetic, pneumatic or electronic in operation.

Mechanical

A car speedometer and mileometer are mechanical counting systems. They are operated by a rotating cable that is driven from the car's gear box.

Electromagnetic

These counting systems use the same principle as a solenoid. Each time the electrical current supplying the counter is switched on and off, one is added to the counter. This could be used to count the number of people entering a museum.
A microswitch would be attached to the entrance door. More information about electronic counters can be found on pages 100 and 101.

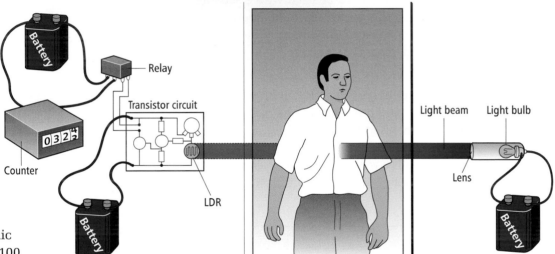

Pneumatic

A pneumatic counter will count pulses of air. It is similar to an electromagnetic counter but is safe to use in dangerous situations, such as where explosive substances are used, as it does not create any sparks (which could happen when using an electrical device).

A pneumatic system using an air bleed could be used to drive a counter. Each time the air bleed is blocked or its tube squeezed, the counter advances one.

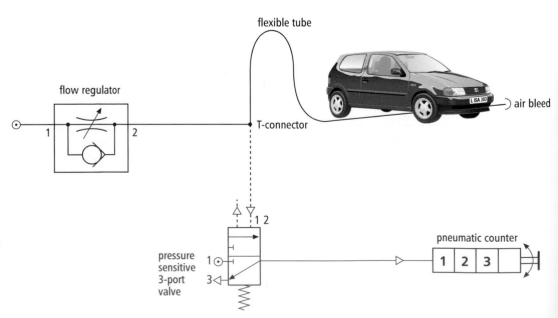

Keeping it Safe

It is important that while your product is in use there is a protective guard around the moving parts. This will stop anyone getting hurt by the device or if something breaks during the test.

It must not be possible to use the equipment until the guard is fully closed. You can achieve this by putting a switch, which is operated by the guard, in series with the power supply. The switch can be operated electrically or pneumatically.

Ask your teacher to show you how microswitches are used on the machines in the school workshop, to ensure they are not started without a guard in place.

This same principle is also used in industry where large machines are used, such as steel presses in car manufacture. An infrared beam is used to detect if the operator is near the machine. If the beam is broken, the machine will not operate.

It is also important to ensure the exercise machine cannot be damaged, which could happen if it becomes jammed.

You could use a belt drive as part of the system. This will slip if the system becomes overloaded. More information about belt drives can be found on pages 128 and 129.

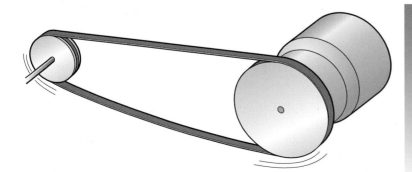

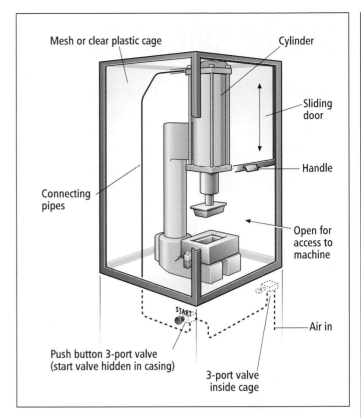

The machine above will not operate until the door is closed, which depresses the 3-port valve inside the cage and allows the air to flow to the start button and cylinder.

■ ACTIVITY

Use a pneumatic counter and air bleed circuit to make a system to count pupils entering your classroom.

IN YOUR PROJECT

► Design an automatic counter for your system.
► Explain the reasons for your choice.
► How will you ensure the safe operation of your system?

KEY POINTS

● Counters can be used with electrical, mechanical and pneumatic systems.
● Safety is an important consideration in the design of any system.

Counting with a PIC

To find out more about PICs go to:

www.dtonline.org

A PIC (Peripheral Interface Controller) is very useful as it can replace many different components and increase the function of a circuit. This means circuits will use less components and their operation can be modified quickly by changing the PICs program.

Using a PIC in your Project

There are many different PICs available. A particularly useful one is the 16F84. This has eight output and five input pins. It must be correctly connected to work reliably.

The PIC requires a stabilised power supply of 6 volts and a 4MHz resonator to control the PICs internal clock. A reset switch can be added which will allow the PIC program to be restarted.

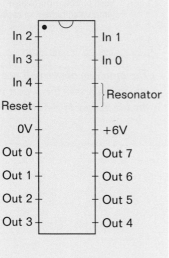

Programming your PIC

There are a several commercially available systems suitable for use in school. It is best to draw a flow chart for your system before you start and map out how each of the pins will be used. You can then program the PIC and use a project board to check its operation before you build your own Printed Circuit Board (PCB). Two program styles can be used, either a flow chart that uses symbols or a list of instructions to be followed by the PIC. Each line must have its own identity code. Some systems need a computer, while others are standalone.

Input Switch

It is possible to use a PIC to count. The number of inputs can be displayed by using a variety of different LEDs and other types of outputs can be turned on after a certain number has been reached.

A digital input signal can be sent to a PIC by using a switch, such as a micro-switch or reed switch. These need to be connected to the PIC as shown below. A micro-switch could be triggered by a cam, with a small rise.

A coursework project that uses two different technologies would potentially achieve a high grade.

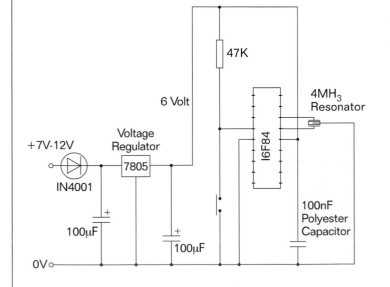

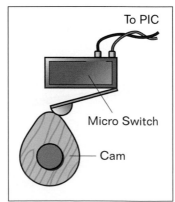

Output Devices

The output current from a PIC is quite small and will only switch on a LED. If you want to control a higher current output device, such as a motor or buzzer, then you will need to use a transistor switch. You can control the direction of the motor by using a second output pin. The simplest way to achieve this is to use a motor driver IC such as the L293D.

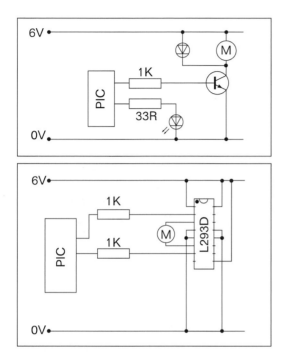

Multi-LED Arrays

Instead of using separate LEDs it is possible to use a multi-array. Ten LED arrays are linked in one package and can be used to form a neat bar-graph display.

Many signs are made up of a series of lights or LEDs

The Seven-segment Display

The Seven-segment display uses seven small LEDs. When these are switched on in the correct sequence they form the decimal numbers 0 to 9, as well as letters. It is possible to use seven of the OUTPUT pins to control the display. You could use a Macro or Sub-routine to switch the correct LEDs on for each number or letter.

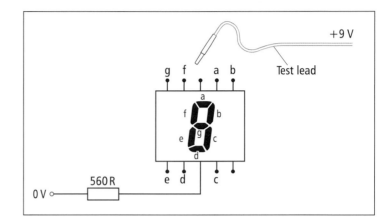

■ ACTIVITIES

1. Write a program to control a series of eight LEDs so they form an interesting visual display.

2. Write a program that could be used with a PIC that would control a pedestrian crossing.

3. Use a component catalogue to identify other PICs that have a different range of INPUTS and OUTPUTS.

4. Discover how a seven-segment display works by lighting each LED in turn. Make the circuit shown on the left using a prototype board. Use the test lead to make each LED light up.

5. Programme a PIC to count. When it reaches ten, switch on a buzzer for 2 seconds.

IN YOUR PROJECT

The clients in the hotel gym will need to keep count of the number of steps they make on the exercise machines. Design a counter and display circuit that could do this.

KEY POINTS

- A PIC can be used to count a number of INPUT pulses.
- A transistor switch must be used with high current devices such as motors or buzzers.
- Seven-segment displays use small LEDs to display a digital number

Keeping Time

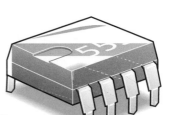

Many systems involve some type of time keeping. By using microelectronics it is easy to build accurate and cheap timing circuits. The 555 IC can be used to do this.

ICT →

You could use a spreadsheet to design a simple frequency calculator.

www. →

For more information about circuits, go to:
www.circuitcit.com

The 555 IC

This IC is packaged in an 8-pin DIL case, which looks like a 741 op-amp. You can find out more about the 741 on pages 90 to 91. The 555 IC can be used in two different ways, as a monostable or an astable timer.

A **monostable timer** has one stable state, when its output is zero. It is possible to change this state for a set period of time, so the output is ON.

An **astable timer** continually changes from one state to another, turning ON and OFF.

The length of time the circuit is ON is controlled by the value of a resistor R and capacitor C. Large values of these components will give a long timing period. The 555 IC is not suitable for timing periods of over 10 minutes. This is because leakage of current from the capacitor makes the timer unreliable.

An egg timer is an example of a monostable timer.

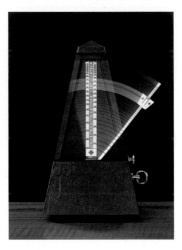

A metronome is an example of an astable timer.

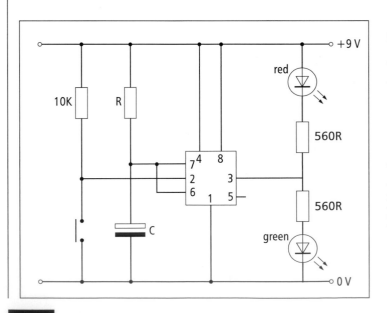

Using the 555 IC as a Monostable Timer

The length of time the circuit is ON can be calculated by using the formula:

time $= 1.1R \text{ (ohms)} \times C \text{ (farads)}$
    $= 1.1 \times 10\,000 \times 0.002$
    $= 22 \text{ seconds}$

When the push-to-make switch is depressed the timing period will start. The red LED will go out and the green LED will come on. At the end of the timing period the LEDs will return to their original state. The timing period can be changed by adjusting the value of R.

Using the 555 IC as an Astable Timer

The IC can be used to produce a series of equal pulses. The length of time for one pulse is called the duration of the pulse, *T*. The ratio of the ON and OFF time is called the mark/space ratio. A circuit can be made with an equal mark/space ratio. The duration of the pulse can be calculated:

$$T \text{ (seconds)} = 1.4 \times R \text{ (ohms)} \times C \text{ (farads)}$$

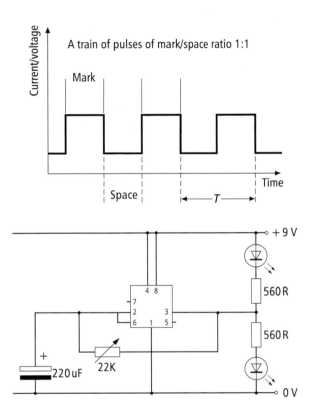

A 555 IC can also be used to make a circuit with an unequal mark/space ratio. The ON (mark) time can be calculated:

$$T_1 \text{ (seconds)} = 0.7 \ (R_1 + R_2) \times C$$

The OFF (space) time can be calculated:

$$T_2 \text{ (seconds)} = 0.7 \ R_2 \times C$$

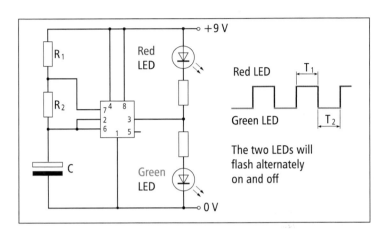

The two LEDs will flash alternately on and off

The **frequency** of the circuit (*f*) is the number of pulses per second. This is measured in hertz (Hz).

The frequency of an astable circuit timer can be calculated:

$$f \text{ (Hz)} = \frac{1.44}{(R_1 + 2R_2) \times C} \quad \left[\begin{array}{l} R = \text{resistance in Mohms} \\ C = \text{capacitance in } \mu \text{ farads} \end{array} \right]$$

Outputs from a 555 IC

The 555 output current is from pin 3. This will switch LEDs on and off. However, if you wish to control a relay or some buzzers you will need to use a transistor (see below). The small current from the IC will switch the transistor on and control the output device.

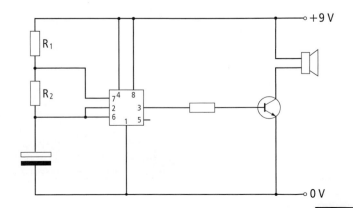

■ ACTIVITY

Calculate the values of a variable resistor and capacitor you would need for a monostable timer that could be adjusted between 30 and 90 seconds.

IN YOUR PROJECT

Clients of the hotel gym need to exercise for set periods of one minute. Design a timer that they could use to inform them when the minute is up.

KEY POINTS

● The 555 IC can be used in timing circuits.
● There are two types of timing circuits, monostable and astable.
● The time period is controlled by the value of a resistor and capacitor.

Designing and Making an Electronic Circuit

After you have designed your circuit you will need to assemble it. There are several different methods you can use. A prototype board will allow you to test the circuit's operation before you solder it together.

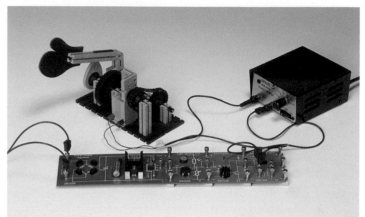

System Boards

These boards can be assembled to model a circuit's operation. They can be used together with construction kits to model a whole system quickly. The effect of any changes on the system can easily be evaluated.

Prototype Boards

These are sometimes called **breadboards**. Components can be fitted to the board and the circuit tested. You can experiment with the circuit design and any changes can soon be evaluated.

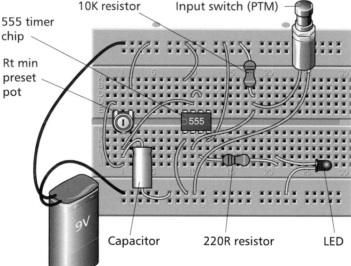

10K resistor   Input switch (PTM)

555 timer chip

Rt min preset pot

555

9V

Capacitor   220R resistor   LED

CAD for PCBs packages enable you to design circuits and to simplify them automatically on screen, and then to print the PCB mask out directly.

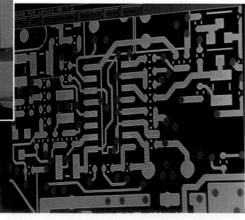

Printed Circuit Boards

Using a PCB will enable you to make a very compact and robust circuit. This method of manufacture lends itself to mass production and is why many electronic products are now inexpensive to buy.

Simple circuits can be drawn directly onto copper clad with an etch resistant pen. For more complex circuits a PCB mask is required.

Modern CAD systems can be used to design PCBs. Many are capable of converting a circuit diagram directly into a PCB layout and mask.

Designing the mask

When you start designing the mask, lay the components out on a mask planning sheet. Allow lots of space between each component and draw in any tracks.

Check the layout and, once you are confident it is correct, redraw the layout so it is more compact. It is best to design the layout from the component side of the board and then use some writing to ensure the mask is always used the correct way round. Mark where pin 1 is on any ICs you use, as well as the positive and negative battery connections.

You can then draw the layout on a CAD system or place a transparent sheet over your final layout design, so you can trace the tracks and pads with a permanent pen.

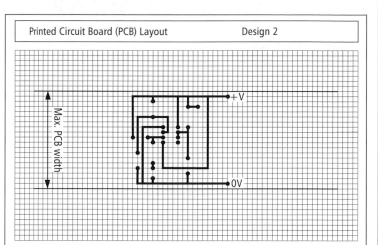

Printed Circuit Board (PCB) Layout   Design 2

Max. PCB width

+V

0V

Your pads and tracks should be as large as possible. This will help when it comes to making your circuit.

A method of making a PCB

1 Produce a mask

Draw the mask design on a piece of clear acetate. This can be done on a CAD system and produced on a laser printer or drawn with a permanent pen.

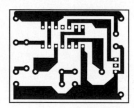

2 Expose the board

Ensure you place the mask the correct way round. Expose the PCB material to UV light for about 2 minutes.

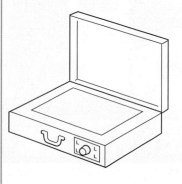

3 Develop the board

Leave the board for approximately one minute in developer solution. The protective layer area that has been exposed to the light will come away. Carefully use a small brush to help this. Once the board has been developed, wash it in water. Always use tongs, gloves and safety glasses when using chemicals.

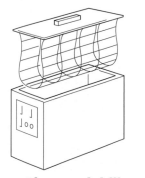

4 Etch the board

Now etch the board in iron(III) chloride or a commercial solution. If the solution is fresh this will take around 30 minutes. The copper that has been exposed should all be dissolved before the board is removed.

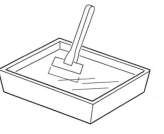

5 Clean and drill

Carefully clean the PCB with some fine wet 'n' dry paper. Do not rub too hard, or you could damage the tracks. The board is now ready for drilling and soldering.

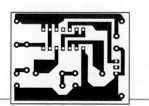

ICT

You could use PCB software to develop your PCB design. Try using the AutoRoute facility as well as your own design.

IN YOUR PROJECT

Test your circuit's operation by assembling it on a prototype board before making on a PCB.

KEY POINTS

- A circuit's operation can be tested by using a prototype board or system board.
- Circuits made on PCB are compact and robust.
- This process is easy to use in mass production.

HI-TECH HOTELS

manufacturing

Assembling, Testing and Evaluating Your Circuit

Once you have designed your circuit and made a PCB, you will need to assemble it. This is done by soldering the components permanently in place on the PCB.

Soldering

It is important to assemble the circuit in the correct order. Start by putting the most robust components in first, such as IC holders and resistors. More sensitive components such as transistors should be left to later. The last components to be fitted should be the ICs.

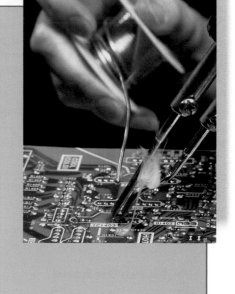

How to solder

1 When soldering it is important to heat both the track and component at the same time, so they become warm.

2 Hold the soldering iron in place for a count of three, then use some solder wire and allow a small amount to melt onto the track.

3 Remove the solder wire.

4 Remove the soldering iron.

5 Your joint should look like a small volcano. If you have not got both the track and component warm enough, a dry joint could occur.

6 Some components, such as a transistor, are sensitive to heat. Use a crocodile clip as a heat sink, to remove the heat so the component is not damaged during soldering.

A good solder joint   A dry joint

Solder   Copper

Component leg   PCB

Testing your circuit

Before you attach a battery, check your circuit. Are all the components correctly connected and joints well soldered? Most problems are caused by poor design or construction. What should you check if your circuit does not work?

First carry out a series of visual checks:

▷ Are the joints well soldered?
▷ Are the components correctly connected?
▷ Are the components of the correct value?
▷ Are there any obvious breaks in the PCB tracks?

Use a multi-meter or Logic probe to check:

▷ If the battery is giving the correct voltage to the circuit.
▷ If there are there any invisible hairline cracks in the PCB tracks.

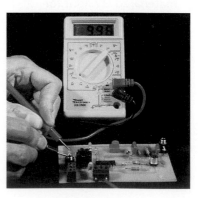

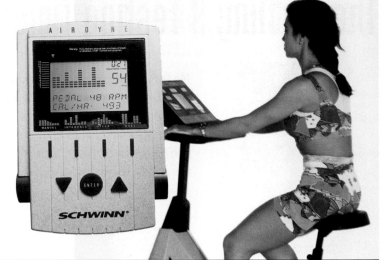

Making It!

Once you have designed your system it will need to be packaged in a case, so it is protected during use. The case will need to be attached to the fitness equipment so it can be removed easily.

All the parts of the system will need to be held securely in the case. You will also need to gain access to the case, so the battery may be replaced.

The case needs to be well finished, and stylish in design, as well as being durable and easy to clean.

Vacuum forming

Vacuum forming is widely used in industry and schools. Make a model of your case from a material such as Styrofoam which can be carved quickly into a variety of shapes.

The vacuum forming mould can be made from any solid material. MDF or Jelutong are ideal as they are easy to work. The mould must have sloping sides so the moulding can be removed afterwards.

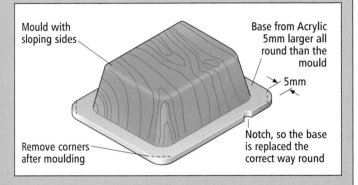

Mould with sloping sides

Base from Acrylic 5mm larger all round than the mould

5mm

Remove corners after moulding

Notch, so the base is replaced the correct way round

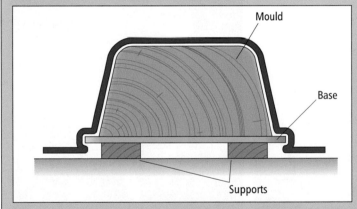

Mould

Base

Supports

Once you have made the mould, make a base for your case. You could use acrylic for this. It needs to be 5mm larger then the bottom size of the mould.

Place the base on a block in the vacuum forming machine and then the mould on top of the base. Once the moulding has been made, it should be trimmed around the bottom edge of the base. Remove the base and then the mould from the moulding. Sand the corners of the base – this will make it easier to remove it from the moulding in the future.

The base should now clip into place in the mould when required.

The base will form a rigid platform, to which the circuit and mounting clips can be attached.

Final Evaluation

Test the operation of your product and check it works reliably. Then show it to other people who may use it. Find someone who regularly uses training equipment. Your PE teacher may be able to help you. What do they think of your idea?

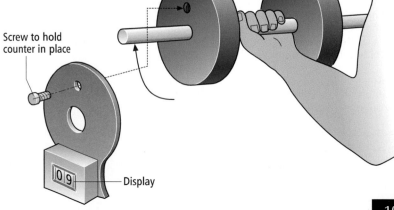

Screw to hold counter in place

Display

HI-TECH HOTELS

final outcome

Theme Two: A Testing Time (1)

Before any manufacturer puts a product into production it will carry out extensive tests to ensure it is reliable and safe to use. Many products will have to pass set industry standards before they can be sold. These may be laid down by the British Standards Institute.

To find out more about JCB go to:
www.jcb.co.uk

Manufacturers are very concerned about the bad publicity they would receive if one of their products proved to be unsafe. They may also want to gain a reputation for making a quality product that will last a long time. It is very easy to get a bad reputation.

There are independent companies, such as the British Textile Technology Group, BTTG, which specialise in testing products and can provide a service to industry.

Magazines like *Which?* and TV programmes like the BBC's *Watch Dog* have helped raise the standards of products now being produced. Customers are also being more critical of what they buy. Just a cheap price is not always good enough. They usually look for a good quality product that lasts a long time.

Case Study

BTTG

The British Textile Technology Group provide a testing facility for the textiles industry. They have two major test houses, one in Leeds and the other in Manchester. They can also carry out specialist fire testing at BTTG Fire Technology Services in Altrincham.

Most materials are tested using the Martindale Abrasion Test. The material under test is stretched over a foam pad by a standard hanging weight. The material is rubbed in a particular pattern and this is compared to a woollen fabric. After every thousand rubs the face of the fabric is checked until two threads are broken.

This provides an accurate comparative test for the wear properties of different fabrics.

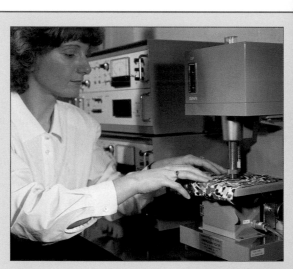

BTTG's uniquely comprehensive range of physical tests covers all stages of textile production together with end-use performance in any of the increasing number of applications for textiles from apparel to agriculture, furnishing to filtration, carpets to civil engineering.

Case Study

JCB

JCB was founded in 1945 by Joseph Cyril Bamford, who bought a second-hand welding kit and made his first product, a farm trailer, in a lock-up garage in Uttoxeter, Staffordshire. Since then the company has grown to be one of the largest producers of construction machinery in the world and is still a family owned firm, run by Joe Bamford's son, Sir Anthony Bamford.

The company has gained a reputation of producing a wide range of machinery, that is of high quality and very safe to use.

Before any of JCB's machines are sold they undergo extensive testing. Any safety feature must be 100% reliable. The company has its own design and research centre and its own test site where development machines can be put through their paces. For example, there is a steel post securely mounted to the floor – a machine's arm can be anchored to this and then continually operated to check its strength and reliability.

A manufacturer must ensure all parts of its product are up to standard, as the smallest fault can destroy a company's good name. JCB buys many parts from outside suppliers and sub-contractors. These must be tested to the same rigorous standards, from switches to hydraulic cylinders. Continual product development and ensuring the highest quality is why JCB has an excellent reputation in the market place.

The safe operation of the machines is one of the most important design features and JCB has been a market leader in improving the quality of construction equipment with several new products and features. These include the new 'Teletruk' and the small 'Robot' range of machines which allow the operator to gain access to the machine from the side, rather than by climbing over the excavator arm. The machine will not operate until the driver is safely in the seat and the safety bar is pulled back over the legs.

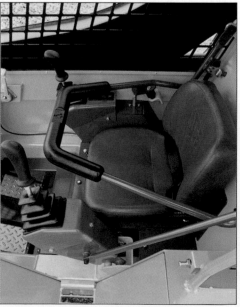

Theme Two: A Testing Time (2)

The motor industry works in a very competitive marketplace. Customers are expecting more performance and reliability from their cars, as well as expecting them to cost less. Manufacturers must rise to this challenge, as well as making cars that comply with all the required safety and environmental regulations.

Testing a Motor Car

Type approval

Before any car can be sold it must pass **type approval**. This means it must undergo a series of rigorous tests to ensure it is safe to drive, as well as complying with current standards relating to environmental concerns about pollution. All aspects of the design and construction will be investigated, and finally an example of the car will be tested to destruction by crashing it into a wall.

Once a car is more than three years old it must undergo an annual MOT test, to ensure it is still safe. Its brakes, lights, exhaust system and safety belts are just some of the items that will be checked. If any of these fail to reach the required standard then they must be repaired before a new MOT certificate can be issued.

Test Facilities

All major car manufacturers have their own test facilities. Vauxhall use the Millbrook test track in Bedfordshire. Other manufacturers may hire this for some tests, and in the past it has been used as part of the RAC rally. Ferrari and Lotus both have their own test race track, so they can evaluate the performance of their cars in secret.

Manufacturers are very concerned about keeping their new designs from one another and therefore it is difficult to obtain access to these sites while a new model is being tested.

Manufacturers use motorsports to advertise their cars and also to test new components and design features

Case Study

Gaydon: Rover Group's Design and Engineering Centre

During the early 1990s Rover Group decided to combine all their design and development facilities on one site. The site chosen was the previous RAF airfield at Gaydon in Warwickshire. The old runways now form part of the 58 km test track. This includes primitive roads and off-road areas of high speed track and many types of obstacles which reproduce a multitude of extreme conditions found in all areas of the world.

During 1993 to 1995 extensive building took place, which included a new 'semi-anechoic chamber' for testing the noise level of cars, a new safety testing facility, new electro-hydraulic test rig facilities and a new Vehicle Operations workshop. By the end of 1996 when the building was completed over 2000 people were working at Gaydon.

Every feature of a new car is extensively tested to ensure that when it finally reaches the customer it works without fault for many years. This will include how well the seat fabric wears and resists fading because of sunlight. All the switches, controls and suspension will receive a lifetime's wear in a matter of days by the use of an automatic test rig.

The electro-hydraulic laboratories will simulate virtually any road conditions. These are referred to as 'shaker rigs' and will dynamically load a test vehicle's suspension in all directions.

The new Combined Road and Environmental Simulation Test (CREST) simulates any road surface and climatic condition.

The temperature can be varied between −40 and +80°C and the humidity between 10 and 95%.

The final test will be to drive a vehicle continuously around the test track. On-board equipment will sense and record the car's reactions. This information is then used to help engineers develop and improve the car's performance and reliability.

Project Four: Starting Point

Quality Counts
(page 114)

Joining
Constructional
Materials
(page 126)

Safety helmets are essential for cyclists, especially on busy roads. Manufacturers need to test the design of the helmets before they can be sold, to ensure they are strong enough.

Can you design and make a device which will enable a manufacturer to test the effectiveness of a cycling helmet?

Get Ahead – Get a Helmet

There has been a rapid increase in the number of people cycling for pleasure in recent years. Cyclists are being made aware of the need to protect themselves by wearing a helmet. These helmets need to be designed so they are comfortable to wear and protect the rider's head, in case of an accident.

In the past, racing cyclists were the only people to wear helmets, and these were of a simple strip design.

With the development of modern light-weight composite materials, attractive and streamlined helmets are now available.

Before these helmets can be sold to the public they need to be tested, to ensure they are strong enough. Although it is possible to use computers to calculate the strength of the helmets, there must still be an impact test.

The Task

Retro Cycling Accessories is a company that manufactures a range of cycling accessories such as bicycle lights and locks. RCA has now started to produce cycle helmets. Before it can sell the helmets they will need to be tested.

RCA have asked you to design and make a prototype test rig that can be used to simulate a cycling helmet being hit, as if it were involved in an accident.

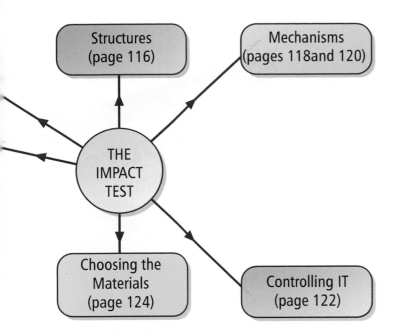

Structures
(page 116)

Mechanisms
(pages 118 and 120)

THE
IMPACT
TEST

Choosing the
Materials
(page 124)

Controlling IT
(page 122)

Before you start your project you will need to consider the following areas.

Investigation

Are there organisations you could contact for help, such as the Royal Society for the Prevention of Accidents (RSPOA) or your local cycling club?

Have there been any test reports about different types of helmets in cycling magazines or *Which?* magazine?

Are there certain regulations about how the helmet should be tested?

Holding the helmet

You will need to design a device to hold the helmet firmly during the test. Does this need to be in the shape of the cyclist's head?

Position of the impact on the helmet

Which part of the helmet is most likely to be hit during an accident? This is where you will need to test the helmet.

Size of the impact

What size and type of force will represent the helmet being hit? Could you calculate this?

Creating the motion of the impact

You will need some form of mechanism to provide the striking force. What type could you use?

Providing the source of power

How will you drive the mechanism, e.g. electric motor, pneumatics, hydraulics?

Testing other products

Many other products need to withstand shock loading. Accidents can be caused by products that are not well designed. An example of this is in toy design. The safety laws on toys in Europe are demanding. All toys need to meet the requirements of British Standard BS 5665 and European Standard EN71.

Instead of a cycle helmet you could look at how a test rig could be designed to test the durability of certain toys or other products.

Release mechanism

How will you start and stop the device?

Recording the results

You will need to monitor what happens to the helmet when it is struck and be able to identify any damage. Will you test the helmet over a number of impacts?

How could you present your results?

A toy is dropped from 850 mm onto an impact plate five times to see if the drive mechanism is exposed by damage (BS5665).

Quality Counts

Manufacturers need to ensure that all their products are of an acceptable standard. A range of techniques has been developed to help check and maintain quality throughout the production process. Standards for manufacture and testing are written down so they are consistently applied by all companies.

British Standards

The British Standards Institute (BSI) was the first national standards body in the world. Its main purpose is to draw up voluntary standards to be observed, and it produces documents which clarify the essential technical requirements for a product, material or process to be fit for purpose.

There are over 10 000 **British Standards** for almost every industry from food to building construction, and textiles to toys. They cover all aspects of production from materials to management.

Certification that a product manufacturing or management process conforms to a stated British Standard provides assurance that an acceptable quality can be expected. This greatly reduces the risk of buying goods and services which could be defective in some way.

Consumer Testing

Many magazines, newspapers and TV programmes carry out comparative tests on a range of products. Most special interest magazines will include a test feature in every issue.

Which?

Which? magazine is published every month by the Consumers' Association, who are a registered charity specialising in the independent testing of a wide range of consumer products. These products include everyday items such as vacuum cleaners as well as services, such as banks or holiday companies. Their tests include interviewing the public and carrying out their own tests.

Ask your teacher if the school subscribes to *Which?* magazine. If it does not, look in the reference section of your local library.

■ ACTIVITY

Plan an article for a cycling magazine, explaining how you would carry out a series of comparative tests on a range of helmets. What features would you look at and how would you test these?

Protecting your Ideas

If you do come up with a good idea then it would not be fair for someone to copy it. You can stop this happening by applying to the Patent Office for ownership of your idea. This means that no-one can copy or use your idea without first getting your permission and then paying you for the right to use it. This is called a **patent.**

The 1977 Patent Act says that, to be patentable, an invention must:

▷ be new
▷ involve an inventive step
▷ be capable of industrial application.

Unpatentable things include those which are:

▷ frivolous
▷ contrary to law or morality
▷ foods or medicines made from known ingredients
▷ words.

Many companies and individuals are very concerned about protecting their design ideas and their image. These range from manufacturing equipment to soft drinks to clothing and pop groups. There have been several high profile cases where companies have taken others to court, to stop them using a product or image which is similar to their own.

Companies register their company logo as a **trade mark**, which protects it from use by others. Material such as music, books and computer software is protected by **copyright**.

There is an international patent office, called the World Intellectual Property Organization, which is based in Switzerland.

In 1933, Percy Shaw came up with the idea of using glass reflectors which could be sunk into the the centre of the road to help indicate traffic lanes. He patented the idea and started to manufacture 'cats' eyes' in quantity, and made a fortune.

WWW. ➡

To find out more about safety standards go to:
www.bsi.org.uk/education

■ ACTIVITY

Find some examples of registered trade marks.

IN YOUR PROJECT

Imagine you wanted to patent your design for your test rig.
Write a description and prepare a series of drawings of it which explain how it works and how it meets the requirements for being patented.

KEY POINTS

● The BSI sets standards for the manufacture and management of a wide variety of systems and products.
● The media help us make choices about the products we buy, by carrying out comparative tests.
● You can protect your idea by applying for a patent.
● Companies and famous people protect their image by using a registered trade mark or copyright.

Structures

Structures are all around us in our everyday lives. They can be man-made, such as bridges and buildings, or can occur naturally, such as eggs and trees.

Structures can be described as either **shell** or **framework**. Shell structures are made of one piece, such as an egg shell. Frameworks are made up of several separate pieces joined together. These individual pieces are called **members**.

A static load

A dynamic load

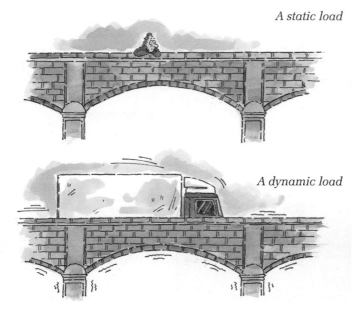

Types of Loads

Structures are designed to withstand forces applied to them, and must not collapse when they are being used. These forces are called **loads**. These loads can be **static** (e.g. the weight of the bridge itself) or **dynamic** caused by movement (e.g. a lorry drives over the bridge).

A compressive force

Compression and Tension

When a structure has a load applied to it each member must withstand this, by either pulling or pushing. A member that has to push away a load is in **compression**. A member that has to pull a load is in **tension**.

A tensile force

Frameworks

Framework structures are easy to make by joining straight lengths of material together. These structures are made up of lots of triangles to make them rigid and strong. This is called **triangulation**. Look at the photograph of the pylon. How many triangles can you see?

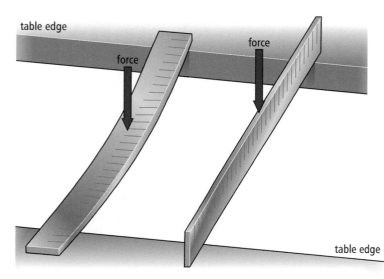

table edge

force

force

table edge

Material Section

The cross-sectional shape of the material in the framework can effect the strength of the structure. A bicycle frame is made from round tubing, as this can withstand loading in all directions.

The amount a beam will bend will also depend on the way it is used. A solid rectangular section is strongest when used with its long side acting against the load. Try this with your ruler.

■ ACTIVITY

Find eight different structures at school or on the way home. Draw them and explain what type of structure they are. When each one is loaded, which members are in compression and which are in tension?

You can find out more about structural design on pages 92–95.

IN YOUR PROJECT

Design a framework structure to support the helmet and mechanism for your test rig.

KEY POINTS

- Structures can be man-made or occur naturally.
- They can be made from pieces of material that are bent or moulded. These are called shell structures.
- Structures made up of several parts or members joined together are called frameworks.
- If a force is trying to squash a structure, the structure is in compression.
- If a force is trying to pull a structure, the structure is in tension.
- The cross-sectional shape of members in a structure can have an effect on the strength of the structure.

Mechanisms (1)

We all use mechanisms in our daily lives. Simple things like door handles and locks can contain quite complex mechanisms. All mechanisms use a force to provide motion. Different types of mechanism can convert the type and direction of the motion. This can be achieved using a cam and follower.

Types of Motion

There are four different types of motion:

▷ rotary
▷ linear
▷ reciprocating
▷ oscillating.

More details about the different types of motion can be found on page 62.

Linear Cams

Linear cams can be used to change the direction of a reciprocating motion. As the cam moves forwards and backwards, the follower will rise and fall as shown below.

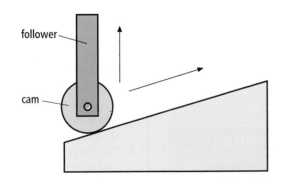

Rotary Cams

Rotary cams can be used to convert a rotary motion into a linear motion. The shape of the cam will control the distance and speed of the linear motion.

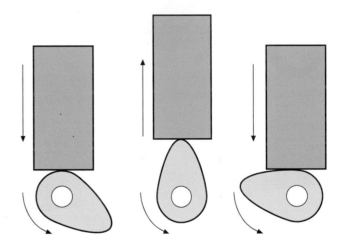

How does the rotary cam work?

As the cam is turned by the rotating shaft the follower moves up and down. The follower is held against the cam by a spring or its own weight. The distance the follower moves is called the lift or stroke of the cam.

Types of Cam

Eccentric or off-set cam

This cam provides a smooth continuous movement, that raises and lowers the follower, known as **simple harmonic motion**.

Pear

Initially, as the cam turns, there will be no linear output movement from this cam. This is called a **dwell**. It will slowly lift and lower the follower.

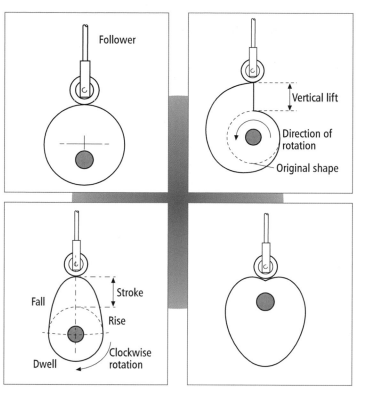

Snail

This cam can be rotated in only one direction. It will slowly lift the follower and then allow it to fall suddenly.

Heart-shaped

This cam is shaped to provide a range of movements for special applications.

Types of Follower

The simplest form of follower is a flat follower. However this will create friction, which can be overcome by using a roller follower. These are more expensive but provide the most hard-wearing type of follower.

If the cam is a complex shape, then you may need to use a knife follower, which provides the most accurate conversion of movement.

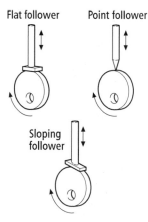
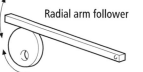

■ ACTIVITY

A car engine uses cams to operate the valves that let the petrol into the cylinder and the exhaust fumes out. Find out how this system works and make a drawing of it.

IN YOUR PROJECT

▶ What type of movement would you need to provide the striking force in your system?
▶ How might you use a cam to achieve this?
▶ Draw a cam and follower mechanism, showing how this would work.

KEY POINTS

- Linear cams are used to convert the direction of reciprocating motion.
- Rotary cams convert a rotary motion to a reciprocating motion.
- The shape of the cam will control the speed and distance of the output motion.
- The type of follower can be changed to suit the shape and application of the cam.

A TESTING TIME

mechanisms

Mechanisms (2)

You can use a mechanism to change one type of motion to another type of motion. Reciprocating motion moves backwards and forwards. Cranks and sliders can be used to convert rotary motion to reciprocating motion, or vice versa.

Cranks

The simplest form of crank is a crank handle. Increasing the distance between the axle and handle will increase the turning force on the axle.

A bicycle has two cranks, which have the pedals on them. These convert the force from the cyclist's legs into a rotary motion. This then turns the rear wheel of the bicycle, by using a chain and sprocket drive.

A car engine uses a crank shaft to convert a reciprocating motion into a rotary motion. This is called a crank and slider and is also found on a steam engine.

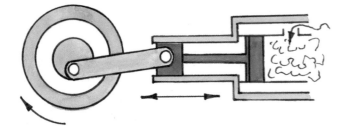

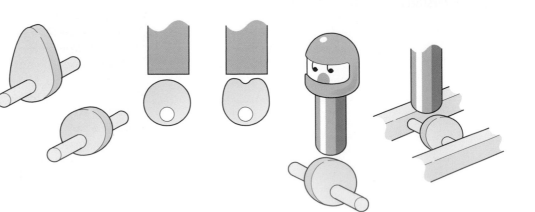

Cams

Cams can be used to convert rotary motion to reciprocating motion. More details about cams can be found on pages 118 and 119.

Rack and Pinion

A rack and pinion can be used to convert a rotary motion to a linear motion. The gear wheel, called a pinion, meshes with a straight bar with gear teeth on it. This is called the rack.

If the gear is rotated in a fixed position the rack will move in a linear motion.

Friction

All mechanisms have **friction** in them as the surfaces rub together. This will reduce the efficiency of the mechanism and cause the parts to wear.

To stop this happening, moving parts can be lubricated with oil and rotating shafts supported in a plain bearing or bush. These bearings can be made from a self-lubricating plastic such as nylon. They are cheap, quiet running, easy to make and when they become worn can soon be replaced. Metal bearings can take greater loads. They are made from materials such as brass or bronze, but these will need some form of lubrication.

Ball bearings, as found in cars and bicycles, are very free-running and will take high loads. If lubricated they will last a long time.

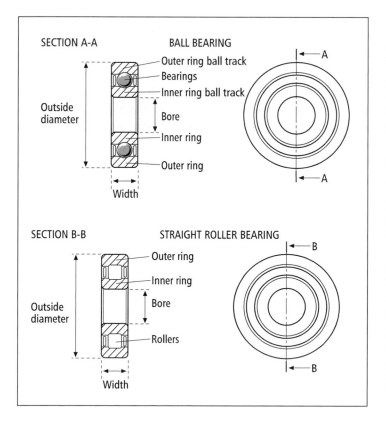

SECTION A-A   BALL BEARING

- Outer ring ball track
- Bearings
- Inner ring ball track
Outside diameter
- Bore
- Inner ring
- Outer ring
Width
A

SECTION B-B   STRAIGHT ROLLER BEARING

- Outer ring
- Inner ring
Outside diameter
- Bore
- Rollers
Width
B

■ ACTIVITY

Find a drawing or photograph of a bicycle and indicate where bearings have been used. Why have they been used in these places?

IN YOUR PROJECT

► How could you use a crank and slider in your system? Draw your idea and then make a card model of the mechanism, showing how it will work.

► Which parts of the mechanism will be subjected to the greatest loads? How could you use a bearing in your idea?

KEY POINTS

- A crank handle is like an axle with a 90° bend in it.
- The longer the distance between the axle and the handle, the better the leverage.
- A crank and slider can change reciprocating motion to rotary motion and vice versa.
- Friction is caused when two surfaces rub together.
- Friction can be reduced by the use of lubrication and bearings.

Controlling IT

Computers are used to control automatically many everyday products. The central part of the computer is the microprocessor. The control systems in microwave cookers, motor cars and video recorders all use microprocessors.

It is easy to change the program in the microprocessor, so a different sequence of operations will be carried out.

ICT

You could use flow diagram software to model your control program.

Computer Control

Most computers found at school and in the home can be used for control. You will need an **interface** box. This will plug into one of the ports at the back of the computer. Ask your teacher to do this for you.

The computer cannot directly control devices like motors and buzzers. These devices can be plugged into the interface and the computer will send a signal to switch them on and off.

It is also possible to plug switches and other sensors into the interface. These can be used to send both **analogue signals** (e.g. dimming a light) and **digital signals** (e.g. on/off) to the computer. Components such as LDRs, thermistors and potentiometers can be used as sensors.

The Program

The advantage of using a computer is that it is very flexible. The program can be changed to suit your requirements. It can be written to carry out a number of operations in a set sequence, or to react to changes in the outside environment, by using sensors. A central heating system uses a microprocessor. It keeps a house at a constant temperature and will switch the heating off when it is not needed.

A PIC would be used in a system like this.

Using the outputs

The interface will switch low voltage motors and buzzers on. Your teacher will be able to tell what voltage the interface will switch. If you need to use a larger voltage then this can be achieved by using a relay.

The interface will switch the relay on, which in turn will switch on the high voltage device. This is how a microprocessor controls a high voltage water heater.

You could use a computer to control your test rig. You may need to use a relay to switch a pneumatic solenoid valve, which controls a pneumatic cylinder. You can find out more about pneumatics on pages 56 to 61.

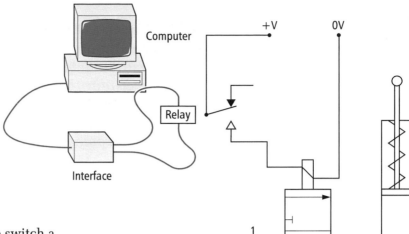

Bar codes

Nearly all products sold have a unique identification bar code on their label or packaging. These can be read by a laser beam which converts them into unique identification numbers.

A central computer belonging to the shop or store is able to monitor sales of each product and help control stock levels. Detailed information about the success of promotional offers can be obtained

quickly. Any price changes can be put into effect immediately by entering them into the system.

How to read bar codes

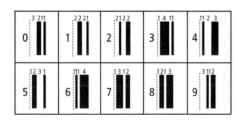

Each number is made up of two black stripes. The stripes can be one of four different widths, as can the spaces, adding up to a total of seven units.

The second, third and fifth numbers are printed backwards, and the black and white stripes are reversed in the sixth to twelfth numbers. This reduces computer error.

Special computer programs exist to create and recognise the bar code pattern.

■ ACTIVITY

Use a computer and interface to control a model automatic car park barrier. Make certain the barrier will not come down until the car has passed through.

IN YOUR PROJECT

What advantages are there in using a computer to control your test rig?

KEY POINTS

- A microprocessor is a flexible way to control systems.
- It can react to input signals.
- It is easy to change the program and sequence of operations.

Choosing the Materials

It is important to make the right decision when deciding what type of material to use when making your test rig. To help you make this choice, carefully consider the factors that will affect your decision. You must look at what properties are required from the material, so it will satisfy your requirements.

You could use a CD-ROM Encyclopaedia to find out more about modern materials.

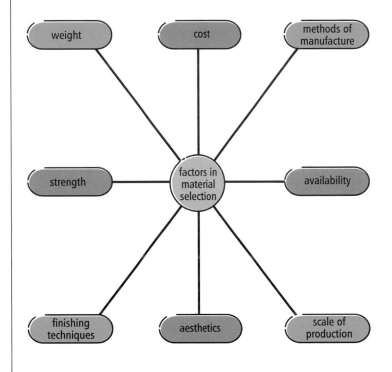

Making the Decision

Each time you choose a piece of material for your project work, ensure you think about each of the factors shown on the left. These factors are inter-related and you will have to balance them so you can make a suitable choice.

Don't choose a material just because it is all you could find or it is what you used last time. Ask your teacher for some help. If you need a different material it may have to be ordered for you. This will take time and must be taken into account when planning your project.

Material Properties

All materials have different properties. Some are very light but may not be strong enough, such as polystyrene. Another may be strong but very expensive, such as carbon fibre. Your final decision will have to take all these factors into consideration.

Material properties of wood

Name	Country of origin	Colour	Characteristics	Properties	Uses
Parana Pine	South America	Pale cream to brown and purple	Tough with a fine grain and few knots	Tends to shrink rapidly on drying and to twist and so is used in jointed constructions	Shelves, cupboards, fitted furniture, step ladders
Mahogany	Central and South America, West Indies, West Africa	Pink to reddish brown	African (Gaboon, Sapele, Utile) quite hard and strong	American easier to work	Indoor work only, furniture, panelling, veneers, pattern making
Beech	Europe, British Isles	White to pale brown, speckled appearance	Hard and strong with close grain, works easily but not durable outside	Good for turning on a lathe	Furniture, toys, tools, kitchen utensils, e.g. rolling pin, spatula, steak mallet

Strength

The strength of a material is its ability to resist having its shape changed. This could be by:

▶ pulling the material and putting it under tension
▶ pushing the material and putting it under compression
▶ twisting the material and putting it under torsion
▶ bending the material
▶ cutting the material: this is called shear.

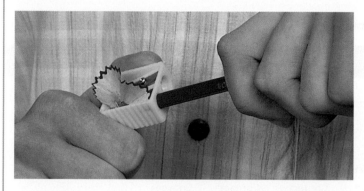

The pencil is under torsion as it is being twisted against the blade in the sharpener.

Stiffness

A material's stiffness is its ability to withstand being bent.

Ductility

A material that is ductile has the ability to be stretched or bent without breaking. Copper is a very ductile metal.

Brittleness

Brittle materials will break very suddenly, as they do not bend. Glass is a brittle material, as is acrylic.

Toughness

A material's toughness is measured by its ability to withstand impact. Ask your teacher to show you the relative toughness of different materials, by holding sample pieces in a vice and hitting them lightly with a hammer.

Hardness

A material is hard if it has the ability not to be scratched.

Modern Materials

In recent years there has been a rapid growth in modern composite materials, such as carbon fibre and Kevlar. These are very strong and light but expensive. Carbon fibre is widely used in motor racing but is also used in boat construction and climbing helmets. Titanium is a very light and strong metal that has been used in military applications for many years, but now can be found in consumer products such as cameras, watches and mountaineering equipment.

In the 1960s a British company, called Speedwell, made titanium bicycle frames. The Spanish rider L. Ocana used one to win a mountain stage of the Tour de France.

Titanium is used in the construction of some modern cameras

■ ACTIVITIES

1. Choose one particular material property and design a simple test rig that could be used to carry out comparative tests on a range of different material types. You could then make this from a construction kit and use it to test some small samples.

2. You may also need to consider other material properties, such as a material's ability to insulate against the cold or an electrical current. It is also important to remember that not all metals are magnetic. Carry out a small test and find out which metals are not magnetic.

IN YOUR PROJECT

▶ List the factors you need to consider in the design of your system, in their order of priority. From this list decide on your material choice and explain your reasons.
▶ Carefully consider the use of your system, before choosing what material it will be made of.

WWW.

To find out more about materials go to:
www.matweb.com

KEY POINTS

● All materials have different properties that make them suitable for a range of applications.

Joining Constructional Materials / Final Evaluation

Materials can be joined together using a range of different methods. Joining materials together is called fabrication. There are three main types of joint. They are permanent, non-permanent and flexible.

Permanent Joints

A permanent joint cannot be taken apart easily. There is now a wide variety of strong adhesives that can be used. In many cases these are stronger than the material they are joining. It is important to make the correct choice of adhesive for the materials you are using.

Materials can also be joined by using heat (welded). Metals and plastics can be welded. This is done by heating the material to its melting point and adding a filler rod or welding rod.

Which Adhesive Should you Use?

Material	Wood	Metal	Acrylic	Melamine	Fabric	Rubber	Expanded polystyrene
Wood	PVA						
Metal	Epoxy resin	Epoxy resin					
Acrylic	Epoxy resin	Epoxy resin	Epoxy resin Tensol				
Melamine	Contact	Contact	Contact	Contact			
Fabric	Contact PVA	Contact	Contact	Contact	Copydex PVA		
Rubber	Contact	Contact	Contact	Contact	Contact	Contact Rubber solution	
Expanded polystyrene	PVA	PVA	PVA	PVA	PVA	PVA	PVA

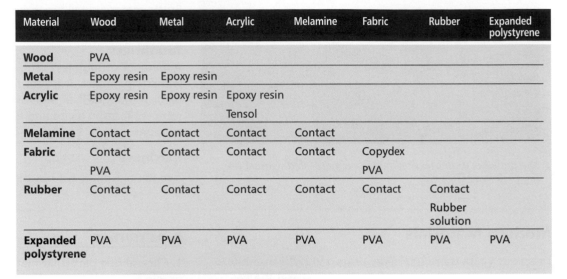

Steel can be brazed together. This is sometimes called hard soldering. This uses a copper–zinc alloy that melts at about 850°C. Flux is first added to the joint and then the steel is heated until it is red hot. The brazing rod is added and melts. Once cool the joint is finished.

Flexible Joints

Flexible and movable joints are widely used in mechanisms. Some of these use one material sliding on another, as in the case of a bicycle chain. In others the joint is made from the flexible properties of the material, such as in food containers.

Non-permanent Joints

Materials can be joined using a range of screws, rivets, nuts and bolts. These joints can be taken apart if required.

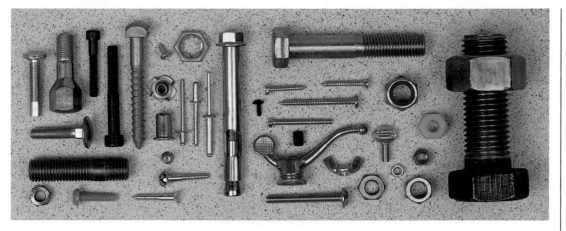

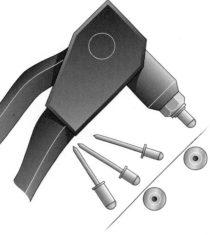

Pop rivets or blind rivets can be used to join sheet materials together. They are especially useful where it is not possible to gain access to both sides of the material, such as joining parts to a tube. They are used to join fittings to aluminium alloy masts on small sailing boats.

Nuts and bolts can provide a strong joint but can become loose. This can be caused by vibration. It is possible to prevent this by using a second lock nut, spring washer or self-locking nut, which has a nylon insert.

Nylon or fibre lock nut

Spring washer

Washer

Test Methods and Results

It is important that all the tests you carry out are fair and consistent. This is so the results can be compared from one product to another.

The simplest way to test the physical strength of a product is to use a rubber hammer as a pendulum and strike the casing a number of times.

You need to design your system so this operation is automated, and the number of strikes is automatically recorded. After 10 strikes the testing rig needs to be stopped so the product under test can be inspected. This could be controlled automatically by using a 4017 decade counter or by using a computer and interface. This will switch the motor off once the counter has reached 10.

Final Evaluation

After using your test rig a few times, check to see it has not suffered any damage. Could the design be improved so it can withstand the shock loading?

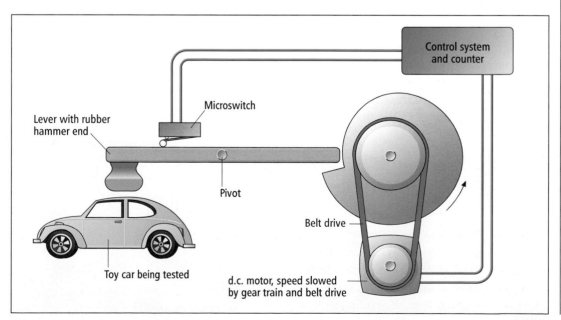

Control system and counter

Microswitch

Lever with rubber hammer end

Pivot

Belt drive

Toy car being tested

d.c. motor, speed slowed by gear train and belt drive

Project Five: Starting Point

Pushing and Pulling
(page 130)

Making It
(page 144)

Manufacturers need to ensure that all the products they are making are to an acceptable quality. A range of tools and techniques has been developed to help check and maintain quality over a long production run. Before a product is put into production, extensive testing is carried out to ensure it will work correctly over its expected life span.

Read through the task on the right. Before you start to develop and finalise your suggestions for a design solution, however, you will need to work through the sections which make up this project and learn more about a range of different electrical and mechanical control systems.

Introduction

A manufacturer will need to simulate the wear and tear a product receives in its life, as part of the product development. By continually operating the product over a short period of time, it will receive a lifetime's use in a fraction of that time.

Motor manufacturers achieve this by driving cars non-stop around a test circuit. This will include all the manoeuvres a car is normally put through.

A manufacturer will also try to ensure that a product can withstand a certain amount of misuse. Although most people would use a screwdriver correctly, at times it is a tool that is put to many different uses, such as opening a tin of paint, then stirring it and finally using the handle as a hammer to replace the lid firmly. All these are quite reasonable uses of a screwdriver and the designer must consider these factors in the development of new products.

Watkins Electrics

WE are a company that manufactures a range of electrical household products. We have decided to break into the popular personal stereo market by designing and making a range of personal stereos aimed at teenagers. Once we have made our first prototype, we want to test its operation over a period of time to check its reliability and durability.

WE would like you to design and make a machine that can continually operate the controls on the personal stereo. In this way we hope to expose the stereo to a lifetime's use, in a short period of time.

128

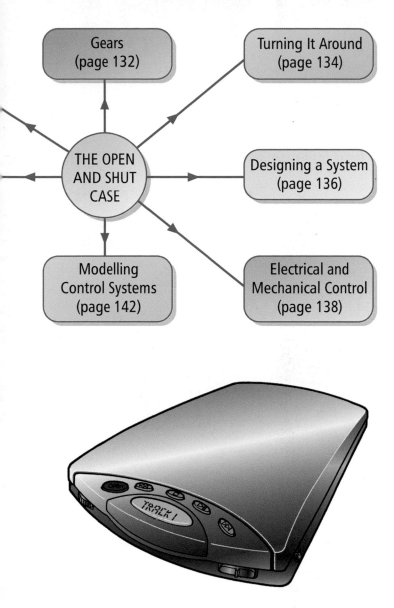

Gears (page 132)

Turning It Around (page 134)

THE OPEN AND SHUT CASE

Designing a System (page 136)

Modelling Control Systems (page 142)

Electrical and Mechanical Control (page 138)

Starting Points

Before you start your project you will need to consider the following. Are there any more you can think of?

Sequence of operations

In what order are the controls normally operated? Are there incorrect sequences that a user could try by mistake? You will need to simulate these. Could you have different or random sequences in your test?

Operating the controls

How are the controls activated? What device could you use to operate these?

Opening and closing the case

Does the case door open by pushing a button or must it be pulled open? How will you close it?

Holding the personal stereo

The stereo will need to be held firmly during the test. But the method of holding must not stop the stereo from being operated.

Safety

It must not be possible for anyone to start the test while their hands are near the machine. What happens if the stereo breaks during the test? Could someone get hurt?

Recording the results

You will need to monitor what happens during the test, identifying points when failure of any component occurs. You may also wish to stop the test at given points to examine the condition of the stereo.

Personal Study

Measure a personal stereo and complete an orthographic drawing of it. Ensure your drawing conforms to BS308. Include the overall dimensions.

Find out how the controls work. Draw each one and show how far it needs to be pushed or rotated for it to be operated. How could you find out what force is needed to operate each control?

Draw a flow diagram to illustrate the sequence of operations your system will need to perform so it tests all the features of the personal stereo.

■ ACTIVITY

Find three products at home that you feel need to be tested before they are sold. Explain the tests that should be carried out and why they are important.

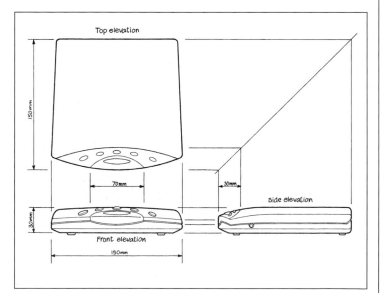

Top elevation

150 mm

70mm

30mm

Side elevation

30mm

Front elevation

150mm

Pushing and Pulling

Levers and linkages can be used to transmit motion and change its direction. They can be found in many everyday objects. It is possible to increase the force applied by a lever or linkage by changing its length.

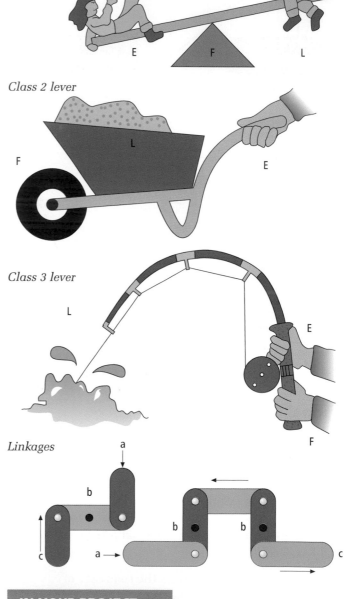

Class 1 lever

Class 2 lever

Class 3 lever

Levers

A lever consists of a **beam**, a **load**, an **effort** and a **fulcrum** (or pivot). There are three different classes of lever, shown on the right. The relative positions of the load (L), effort (E) and fulcrum (F) will determine the class.

The bell crank is a different type of lever. This is used to change the direction of a movement through 90°.

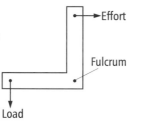

Linkages

It is possible to join two or more levers together to form a linkage. By moving the position of the fulcrum, it is possible to increase the force that is applied by a linkage and the distance it moves. This feature is used to improve the performance of bicycle brakes.

Linkages

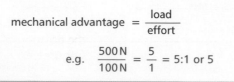

Just a moment

When a force (effort) acts on a lever it makes it rotate. This turning force is called a **moment**. Moments can be calculated to find out how much effort will be needed to move a load. The ratio between the load to be moved and the effort needed is called **mechanical advantage** (force multiplication):

The larger the number, the greater the mechanical advantage. Class 1 and 2 levers give the most advantage. This means that a large load can be moved using a small effort. Class 3 levers are less common because their mechanical advantage is less than 1.

$$\text{mechanical advantage} = \frac{\text{load}}{\text{effort}}$$

$$\text{e.g.} \quad \frac{500\,\text{N}}{100\,\text{N}} = \frac{5}{1} = 5{:}1 \text{ or } 5$$

IN YOUR PROJECT

Moment calculations can be very useful in coursework projects, but will not be tested in the written examination.

130

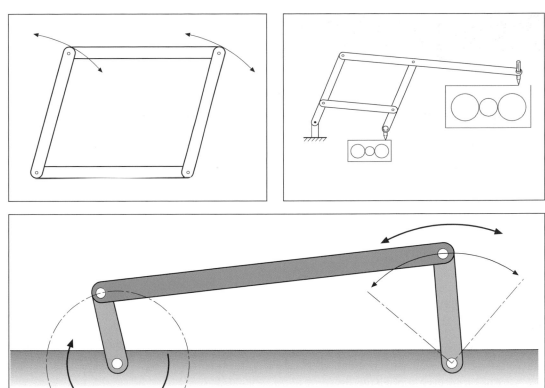

Parallel Linkages

If a linkage requires two parts to stay parallel during its operation, then a parallel linkage can be used.

Treadle Linkage

By combining a crank with a parallel motion, it is possible to convert a rotary motion into an oscillating motion. This is called a treadle linkage and is used on some types of windscreen wipers.

Toggle Clamp

A toggle clamp is a linkage that can be used to hold a piece of material firmly in one place. Mole grips use this principle. A toggle clamp can also be known as an 'over centre' clamp as the middle of the clamp is pushed 'over' its centre to hold the material in place.

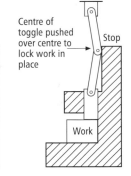

Centre of toggle pushed over centre to lock work in place

Stop

Work

You could animate the motion of a mechanism by using 2D CAD software.

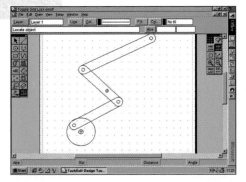

■ ACTIVITY

Find a mole grip, and use a construction kit or soft materials to model how it works.

IN YOUR PROJECT

► How could you use a linkage in the design of your system? Draw it first, and then make a card model to check if it works.
► Could you use a toggle clamp to hold the personal stereo in place while it is tested?

KEY POINTS

● A lever is formed by a beam, with a load, effort and fulcrum.
● Linkages are formed by joining levers together.
● The force and distance moved by a lever or linkage can be changed by the lever's length and by the position of the fulcrum.

Gears

Gears can be used to transmit rotary motion from one shaft to another. It is possible to increase or decrease the rotational speed (velocity) of the output shaft.

Gears can also be used to change the direction of the motion and can transmit large forces.

Types of Gears

Spur gears

The simplest form of gear is called a spur gear. These have teeth which mesh with one another and give a positive drive. This is called a **gear train**. However, spur gears do need to be aligned carefully and lubricated, as in a car's gear box. In applications with a light load, such as a food mixer, the gears may be made from plastic. This will reduce noise and the need for lubrication.

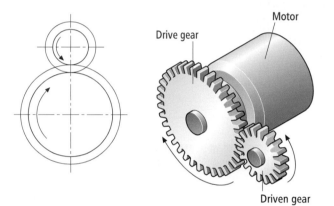

Motor

Drive gear

Driven gear

Try to remember...

A *large driver* gear turning a *smaller driven* gear will *speed up* the velocity of the gear train.

A *small driver* gear turning a *larger driven* gear will *slow down* the velocity of the gear train.

If you want the gears to rotate in the same direction, you need to put an idler gear between them.

Input	Process	Output
Smaller gear turning quickly in a clockwise direction	Gear train	Larger gear turning slowly in an anticlockwise direction

Worm gears

A worm gear looks like a screw thread and can be meshed with a gear called a worm wheel, to provide a large speed reduction. It also turns the drive through 90°. The worm gear always acts as the driver gear and the system cannot be run the other way round.

Bevel gears

These gears are used to turn the drive through 90°. The velocity of the gear train can be increased or decreased, in the same way as with spur gears, by changing the relative sizes of the two gears.

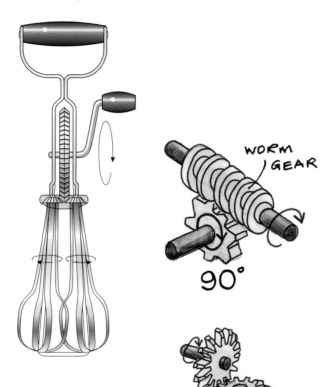

WORM GEAR

90°

90°

BEVEL GEARS

Gear ratio calculations

1. The ratio between the **driver** gear and the **driven** gear, is called the **gear ratio**. This determines the **velocity ratio**.

$$\text{gear ratio} = \frac{\text{number of teeth on the driven gear}}{\text{number of teeth on the driver gear}}$$

If the driver gear (A) has 30 teeth and the driven gear (B) has 90 teeth, what is the system's gear ratio?

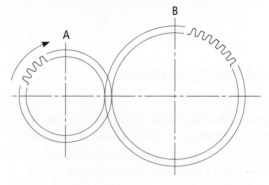

$$\text{gear ratio} = \frac{90 \text{ teeth}}{30 \text{ teeth}} = \frac{9}{3} = \frac{3}{1} = 3{:}1 \text{ or } 3$$

This means the driver gear turns three revolutions, for every one revolution of the driven gear. In other words, it turns three times as fast.

2. If the driver gear turns at a velocity of 150 r.p.m. (revolutions per minute). What is the velocity of the driven gear?

$$\text{velocity ratio} = \frac{\text{driver velocity (r.p.m.)}}{\text{driven velocity (r.p.m.)}}$$

$$\text{driven velocity (r.p.m.)} = \frac{\text{driver velocity (r.p.m.)}}{\text{velocity ratio}} = \frac{150}{3} = 50 \text{ r.p.m.}$$

■ ACTIVITY

Find a product at home that uses a gear train, such as an egg whisk or food mixer. Draw the gear train and explain how it works.

You can find out more about d.c. motors on page 82.

ICT ➡

You could use a spreadsheet to design a simple gear ratio calculator.

IN YOUR PROJECT

▶ Small d.c. electric motors run at very high speeds. How could you use a gear train to slow this down and operate your system?
▶ Use a constructional kit to model the gear train.

KEY POINTS

● Gears can be used to transmit rotary motion.
● Gears can be used to increase or decrease the velocity of the motion.
● Gears can be used to change the direction of the motion.

Turning It Around

Rotary motion can be transmitted using a drive system. This involves the use of pulleys and a belt or a chain and sprockets. The driver and driven pulleys both turn in the same direction.

It is possible to change the speed (velocity) of the system by using different sizes of pulley and sprocket.

Belt and Pulley Systems

A belt and pulley system is a cheap and easy way to transmit rotary motion. The belt must be kept tight, by a jockey or idler wheel, so it does not slip. This wheel can be spring-loaded or adjusted by use of a clamp. The ability of a belt to slip can be put to good use, as it can prevent a machine from being overloaded. A drilling machine uses two cone pulleys, so the speed of the drill may be changed. The belt and pulleys are 'V' section.

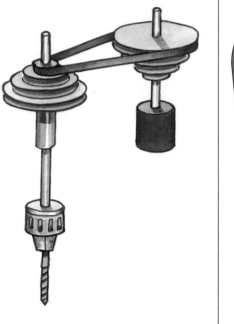

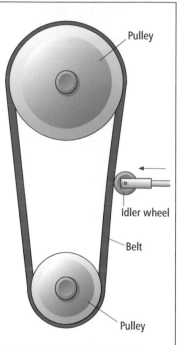

Pulley

Idler wheel

Belt

Pulley

Other types of belt

In some applications it is difficult to fit the belt or to obtain one of the correct length. To help overcome this problem, a linked belt can be used. This can be broken and links added or removed to change its length.

A toothed belt provides a positive, quiet and clean drive system. It is used when it is important that the two pulleys are kept in sequence, such as on a computer plotter or a timing belt on a motor car engine.

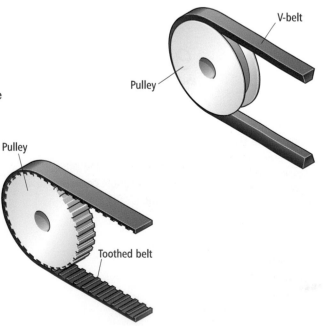

V-belt

Pulley

Pulley

Toothed belt

Chain and Sprocket Systems

A chain and sprocket system is a very efficient way of transmitting rotary motion over a great distance. There is no slip, as can happen when using a belt, and the sprockets do not need to be aligned accurately, as with spur gears. The system must be kept lubricated. The velocity ratio of the system can be changed by varying the size of the sprockets, as used on some bicycles.

■ ACTIVITY

A bicycle uses a chain and sprocket drive. On many bicycles, the gears are changed by using a derailleur.

Draw this system, explaining how it works, its advantages and disadvantages.

IN YOUR PROJECT

Could you use a belt or chain system in your project? Why would the ability of the belt to slip be useful?

KEY POINTS

- Rotary motion can be transmitted over a relatively long distance by using belts and chains.
- The belt or chain must be kept in tension.
- The ability of a belt to slip can stop a machine being overloaded.
- The speed of the system can be adjusted by changing the size of the pulleys or sprockets.

Designing a System

Before designing your system, decide what you want it to do. You should carefully analyse the problem and break it down into simple stages. From this you will be able to draw a system diagram and flow diagram.

Analysing the System

Ask yourself questions about what you need the system to do:

▷ Will the system control a sequence of operations?
▷ Does it have to react to an input signal by using feed back, as in a closed loop system?
▷ What outputs do you need – e.g. motors, relays?
▷ What inputs are needed – e.g. switches, LDRs?
▷ Will the system have to keep count or time? Does this need to be displayed?
▷ Are there any special considerations?

■ ACTIVITY

Design a system diagram for your project to show its major components. Start by drawing a simple diagram.

Build up your system step by step. Add blocks to the diagram, until it is complete and shows your final system.

The diagram below shows a system diagram for a household burglar alarm.

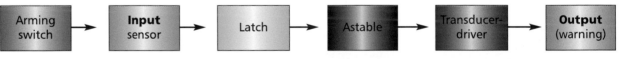

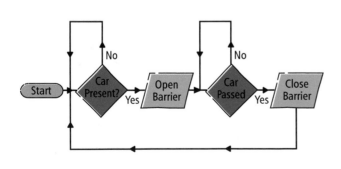

Flow Diagram

The blocks in your system diagram can represent quite complicated operations. To show how these will work you need to draw a flow diagram which shows the system's operation. If your system is very complex with sub-systems, it may be easier to draw separate flow diagrams for each sub-system. Later these can be combined, when you are happy each will individually work.

Once you have drawn your flow diagram you can start to write a computer program. Start by writing a program for a simplified system and check its operation before you add any other features.

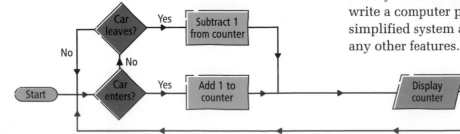

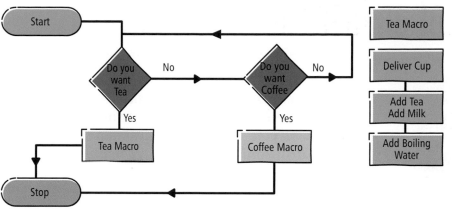

Macros

The operation of a sub-system can be controlled by using a **macro**. The macro will contain a series of set operations. When the program reaches a macro, it will carry out the operations defined in the macro and then proceed to the next step.

Program logic controllers

PLCs are a type of microprocessor that can be programmed to carry out a series of operations. The program is written on a computer and then downloaded to the PLC. If a change of program is required then a new one can be loaded. PLCs are widely used to control machines in the manufacturing industry. If a company wishes to change the component a machine is making, then it is a simple job to change the PLC's program. This offers flexibility in the company's product range.

Safety Systems

Many systems have automatic safety controls or are fitted with a safety alarm. These will only be used in an emergency and must be 100% reliable.

An alarm cord on a train may never be used in a real situation but it must be tested and checked so the rail operator knows it will work when it has to.

Railway signals and trains have complex safety features in their control systems to prevent two trains travelling along the same piece of track. The signals are designed to turn red if any part of the system fails, so all the trains are stopped and the fault can be detected. This is called a fail-safe system.

■ ACTIVITY

Draw a system diagram and flow diagram to show the working of a central heating system which incorporates a timer.

IN YOUR PROJECT

▶ Draw a flow diagram to show the operation of your test rig, so it stops after every 20 cycles.
▶ Explain whether or not your test rig needs any automatic safety features.

KEY POINTS

● The components of a system can be represented by a system diagram.
● The detailed operation of a system can be shown by a flow diagram.
● Macros can be used to control sub-systems.

A TESTING TIME

system design

Electrical and Mechanical Control

It is possible to combine electronic, mechanical and pneumatic systems. You can use a mechanism to control electronic and pneumatic systems, or vice versa. Many systems combine several different technologies to achieve the required output. It is important to understand how technologies can be combined.

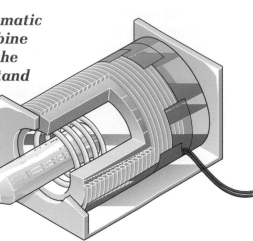

Solenoid symbol

Solenoids

A solenoid uses an electrical current to provide a mechanical movement. It transfers electrical energy into mechanical energy.

Input Electrical energy	→	Process Solenoid	→	Output Mechanical energy

Solenoids are used in electronic combination locks. They provide only a small movement, but this could be amplified by using a lever. More information about levers can be found on page 130.

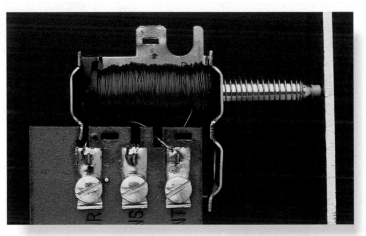

Cam Shafts

By using several cams on one shaft it is possible to control a sequence of operations. A motor car engine uses this principle. The engine's cam shaft controls the correct sequence to open and close the inlet and outlet valves. More information about cams can be found on pages 118 and 119.

The cam followers can be replaced by a pneumatic valve or microswitch. In this way it is possible to control the sequence of operations in pneumatic and electrical systems. The shape of the cam and speed at which it turns will control how long the output device is switched on. The camshaft would be turned by an electric motor that had been slowed down by using gears. More information about gears can be found on pages 132 and 133.

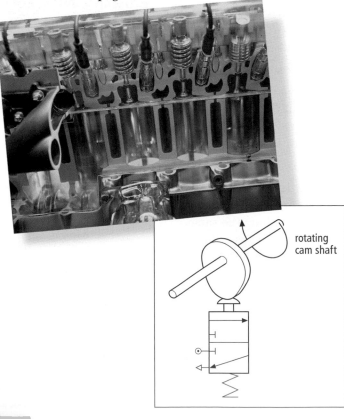

rotating cam shaft

Program Timers

Program timers that are found in washing machines or immersion heaters use the same principle as a camshaft.

The discs in the control system have notches cut in them. As the discs turn they switch a series of microswitches on and off. In turn these control the operations of the washing machine.

Program timers that are used with immersion heaters or security lights work on a similar principle. However, the timing period can be controlled by adjusting small plastic tabs. These can be pulled out, creating the notch that will operate a microswitch.

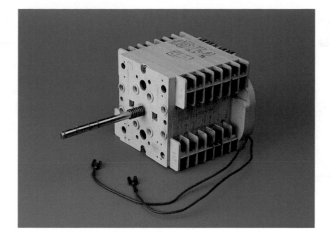

Although an IC could be used to control a washing machine, the heat and moisture inside the washing machine could damage the circuit. An electromechanical control system is more reliable and robust for this type of application.

A disassembled program timer

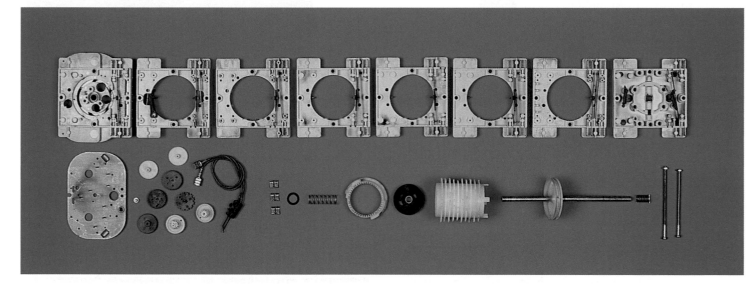

■ ACTIVITIES

1. Use a construction kit and make a model system that uses a cam and microswitch to control an electric motor. Experiment with different shapes of cam and see how they affect the motor's operation.

2. Design a system that uses a cam to operate a pneumatic valve, so it is depressed once every minute.

IN YOUR PROJECT

▶ How many different operations do you need for your testing device?
▶ Draw a flow diagram for its operation and design a camshaft that could control this sequence of operations.

KEY POINTS

● Solenoids transfer electrical energy into mechanical energy.
● Cams can be used to operate pneumatic valves and microswitches.
● A camshaft can be used to control a sequence of operations.

Mechatronics

Many control systems in our daily lives use a combination of microelectronics and mechanisms. This is called mechatronics.

www. ➡

To find out more about robotics go to:
www.schoolnet.ca/ vp-pv/robotics

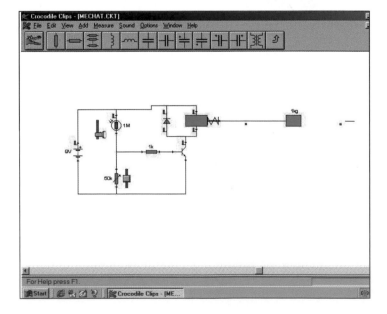

MOSFETs

The output current from a microelectronics system is small. If you want to control a higher current output device, such as a motor or solenoid, then you will need to use a transistor switch.

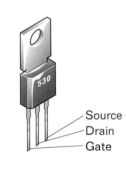

— Source
— Drain
— Gate

If you want to switch a device that requires more than 1 amp, such as a powerful motor or loud siren, then you should use a power MOSFET (a Metal Oxide Semiconductor Field Effect Transistor – see page 78), such as the IRF530, which can switch a large current.

Motors

When using microelectronic systems such as PICs it is best to use higher quality solar motors. These do not create electrical interference, which can disrupt the PIC's operation.

Compound Gears

Compound gear trains use several pairs of meshing spur gears. They are useful to slow down fast revolving motors, as often used in project work, and to accurately control the final output speed. This can be calculated by first finding the Gear Ratio of each pair. See how to do this on page 133. The final gear ratio is calculated by multiplying the gear ratios.

Total Gear Ratio = GR1 x GR2

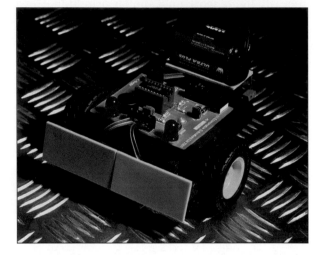

Screw Threads

A screw thread can be used to convert rotary motion into linear motion. They can also be used to apply a large force, as used in a car jack. Screw threads are also used on a centre lathe, in the tail stock and lead screw.

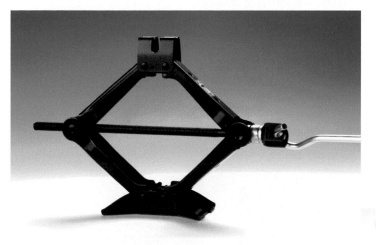

A micro switch can be used to sense when a moving part of a system has reached a desired position, such as when a sliding door is open. This is called a limit switch. These are also used to ensure lift doors are correctly closed before the lift moves.

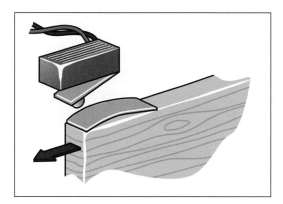

■ ACTIVITY

1. Write a programme for a PIC to control a windscreen wiper, so the wipers will come on intermittently every 5 seconds or continuously.

2. Modify your programme so the intermittent delay time can be adjusted, between three and ten seconds. Add a limit switch so the wipers always stop at the same point.

3. Explore ways a cam and lever mechanism can be combined to amplify the movement of the cam.

Combining Mechanisms

Most systems will combine different mechanisms to achieve the desired outcome. By combining a treadle and parallel linkage you can see how a car's windscreen wipers would work.

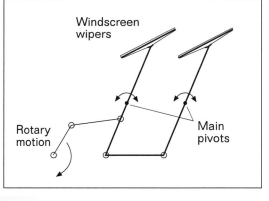

IN YOUR PROJECT

Find out the rotational speed of an electrical motor you may use in your project. You can do this by looking at the motor's data sheet, provided by the manufacturer. This information may be on a CD-ROM or the internet. Design a compound gear train so the final output speed is 60 rpm.

KEY POINTS

- Many control systems use a combination of electronics and mechanisms, this is called mechatronics.
- A MOSFET can be used to switch high current devices such as motors or buzzers.
- Compound gears can be used to change the output speed of rotary systems.
- Simple mechanism can be combined to form more complex ones.

Modelling Control Systems

You should consider modelling your idea before you make it. This will help ensure it will work and save time during the manufacture. There are many ways you can do this, including using specialised computer software.

Modelling in the Car Industry

Car manufacturers will make a large variety of models before a car is put into production. These will include scale and full size models of the car. They will also use computers, which can model in 3D items such as door hinges. The computer model can then be used to ensure the door opens and closes correctly before it is made.

The designer can also use a wire frame computer model of the car to test its structural strength. This stress analysis is called **finite point analysis** and is widely used in mechanical design. It is possible for the computer to simulate a crash. The designer can then modify the design and quickly carry out the test again.

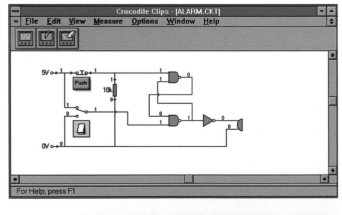

Modelling Electronic Circuits

It is easy to model an electronic circuit or system by using CAD or a kit. Larger complex systems can be tested by using pre-made system boards. You should record and document the use of these in your project folder.

Simpler circuits can also be modelled using commercial systems that use discrete components.

Electronic CAD software can be used to model a circuit and explore its behaviour without actually needing to build the circuit with real components. This will save the designer time and help ensure the circuit will work when it is made.

Some computer packages will also model mechanisms. Another way to do this would be the use of a construction kit. You can find out more about modelling on page 144.

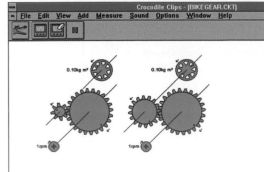

Making it by Computer

It is possible to use computers to control manufacturing equipment. This is called **computer-aided manufacture** (CAM).

Computer numerical control (CNC) machines have been used in industry for many years. However, it is now possible to streamline the whole design and manufacturing process.

Once drawings have been completed using a **computer-aided design** (CAD) system, the information stored in them can be used directly to make the product. This is called **computer integrated manufacture** (CIM).

Using a CIM system helps ensure accuracy, as there is no opportunity for small changes in size to occur because of human error. This is a totally automated production process with a very aspect controlled by computer.

Manufacturing companies which have adopted such systems have been able to make dramatic reductions in their prices and increase their quality and reliability.

Case study:

The AKA mobile phone

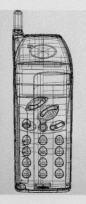

The first stage was to create two appearance proposals for the design of this new portable telephone. These were presented as hand-made models. The models were made from a material called Ureol which is a resin that can be cut and finished easily by hand.

Next AKA were commissioned to combine the designs and the technical specification into a viable product. Alias software was used to create digital models of all the electronic components and to then wrap a 3D 'skin' round them. Buttons were positioned to ensure existing components could be used, saving development time and costs.

Shaded images of the wireframe were constantly assessed to ensure that the model met the aesthetic as well as technical requirements.

AKA then created a physical model to present to the client. This one was made directly from the computer data using a CNC milling machine.

A **rapid prototype** was produced to check that all the electronic components and plastic housing fitted correctly. This was created using a laser cutting a liquid resin, again fully computer controlled from the original computer data.

Fully rendered colour images were then created. Colours, textures and graphics were assigned to the surfaces of the digital model and photo-realistic images created. These pictures were used in brochures and advertising posters for the product launch.

Finally the 3D digital data were sent via a modem directly to a toolmaker in Seoul, Korea, where the production tools were made.

IN YOUR PROJECT

▶ Once you are happy with the operation of your system, look at how it could be modified and used by a person with a mobility handicap to operate a personal stereo.

▶ Could you redesign your system so a handicapped person could use it to control other products?

Making It / Final Testing and Evaluation

When making your system you must carefully consider the moving parts. These must be free running and any mechanism must not jam. Making a working model of your idea and careful planning will help your system to work first time.

Modelling Mechanisms

You can use a construction kit to make a working model of your idea. It is quick and easy to make modifications to ensure your idea will work.

Once you are satisfied your idea will work, draw it accurately. You could do this on a computer aided design (CAD) system. Use your drawings to make a card model of your idea. Use these to check again that your idea will work. The card pieces or computer print-outs can then be used as templates to make the real parts of your system. Using card models is a particularly good way of comparing the movement from different shaped cams. You can find more about cams on pages 118 and 119.

A CAD package can be used to produce templates to help you produce your parts to the correct size and shape.

Height gain of cam

50

30

Axle Position

ø35

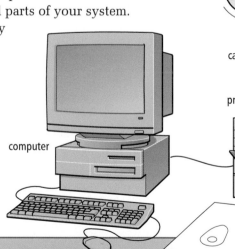

cam profiles being printed out

printer

computer

Working to a tolerance

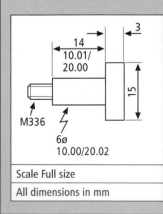

3

14

10.01/ 20.00

15

M336

6ø
10.00/20.02

Scale Full size

All dimensions in mm

All moving parts must have a clearance or gap between each other so they can move. A shaft must be slightly smaller than the hole it goes through so it can turn. In industry, parts are accurately machined to **tolerances**. This is to ensure any shaft, of a given size, will fit correctly when it is assembled with other parts on the production line.

How accurately the part needs to be made will depend on its application. All parts will be made to a **tolerance level**. This is usually expressed by two numbers: an upper and lower limit.

In a simple example, a component intended to be 100 mm in length could vary between 99.1 mm and 100.9 mm. The tolerance is the difference between the upper and lower limits, i.e. 1.8 mm or +/− 0.9 mm.

Quality Control Systems

When a manufacturer is making a large number of components or products, it will set up a **quality control system**. This involves inspecting a sample of the components as they are made and accurately measuring them. By examining the pattern of a series of tests it may be possible to notice if a particular machine is starting to produce components that are getting close to unacceptable tolerance limits. The machine may then be adjusted, so components are not made of the wrong size and have to be scraped.

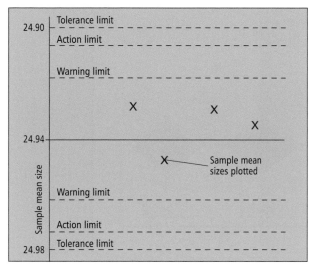

Final Testing and Evaluation

Once you have made your system, run it for a period of time to check it works correctly before you use it to carry out any tests.

▷ **Listen** carefully. Can you find any parts that are hitting or rubbing against one another?

▷ **Watch it** carefully. Are all the movements smooth and continuous?

▷ **Feel it** carefully. If your project does not appear to be running smoothly, disconnect the drive system and operate it by hand. You should be able to feel any points at which it is harder to operate. Are all the parts lined up accurately? Have you have lubricated all the moving parts?

When you are happy with your system's operation then you can use it to carry out some comparative tests.

Ensure these tests are carried out fairly. Always use the same load and number of cycles. You could use a computer or counter to keep a record of this for you. The computer could be programmed to operate the test for a set number of cycles and then stop.

KEY POINTS

● Make a model of any mechanism you are going to use and check its operation.
● CAD can speed up the development of a product, as it is easy to make modifications to a drawing.
● Tolerances are important as they help ensure parts fit together correctly when they are assembled.
● Quality control systems help ensure all parts and products are consistently produced, to the same standard.

A CAD package could be used to produce templates to help you manufacture your parts to the correct size and shape.

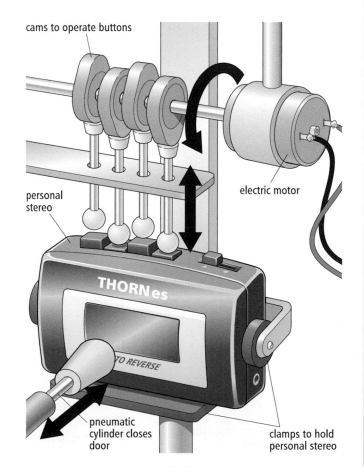

Examination Questions

Your teacher will tell you which of the following questions are appropriate to the focus areas in your course. You will need some A4 and plain A3 paper, basic drawing equipment, and colouring materials. You are reminded of the need for good English and clear presentation in your answers.

1. This question is about using mechanisms. See pages 118-121 and 130-135. *(Total 7 marks)*

A company wants to manufacture a system that will automatically close curtains when it becomes dark.

a)   What is the name used to describe the movement of the motor?   *(1 mark)*

b)   What is the name used to describe how the curtains move?   *(1 mark)*

c)   Name the sensor used in the electronic circuit.   *(1 mark)*

d)   What is the problem with the system as it is shown and how could this be over come?   *(2 marks)*

e)   When used, the curtains move too quickly. Design a mechanical system to slow the speed of movement down.   *(2 marks)*

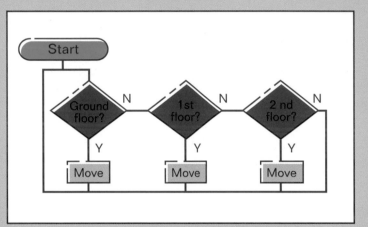

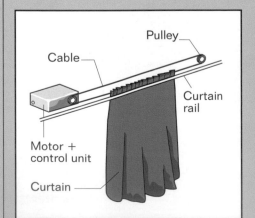

2.   This question is about microprocessor control.
See pages 138-139. *(Total 12 marks)*

The flow diagram above shows part of a system for a lift control in a large shop that has a ground floor, 1st and 2nd floor.

a)   Describe what safety features should be added to the programme.   *(2 marks)*

b)   What term is used to describe the function in a system that stops it working if anything goes wrong?   *(1 mark)*

c)   Show how you would use a sensor to sense when the lift has reached the correct floor.   *(3 marks)*

d)   Use a programming language you understand to design a complete programme for the lift.   *(6 marks)*

3.   This question is about PICs.
See pages 100-101. *(Total 12 marks)*

a)   A PIC has to be used with other components to make it work. Name one of these components and describe what function it performs.   *(2 marks)*

b)   What are the advantages and disadvantages of using a PIC in a product?   *(4 marks)*

c)   Draw the input part of a circuit that could be used with a PIC to sense when a switch is pressed on.   *(2 marks)*

d)   Why can a motor not be directly connected to an output pin of a PIC? Suggest a suitable interface device.   *(4 marks)*

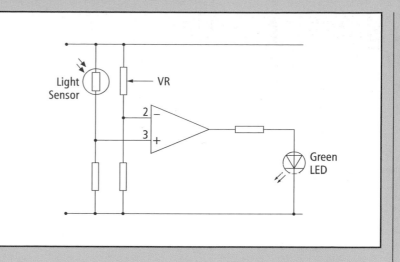

4. This question is about industrial practice.
See pages 142-145. *(Total 13 marks)*

In the past children's toys had been made from lead or tin plate. They are now made from moulded plastic.

a)   Name a suitable process and material that could be used for the manufacture of children's toys.
(2 marks)

b)   Explain two advantages of using this material and process.
(4 marks)

c)   The toy is designed on a CAD system. What do you understand by the term CAD? *(1 mark)*

d)   What health and safety issues must be considered when designing children's toys? *(2 marks)*

e)   What must any manufacturer complete before it markets its products? *(2 marks)*

f)   What do you understand by the terms quality control and patent? *(2 marks)*

5. This question is about sensing.
See pages 96-97. *(Total 16 marks)*

A sporting company wants to design a device to accurately sense when the light level drops below a certain point. A green LED must come on when it is bright and a red LED must come on when it is dark.

a)   What sensing component would you use to sense light?
(1 mark)

b)   The company has decided to use an operational amplifier in the final circuit. This is shown above. Explain how the level of light needed to switch the green LED on can be adjusted. *(2 marks)*

c)   Explain how the op-amp uses the voltages at pins 2 & 3 to control the output voltage. *(4 marks)*

d)   The red LED is missing from the circuit diagram. Add the component and explain your reasons for your answer. *(4 marks)*

e)   Explain how Computer Aided Design (CAD) could be used in the design of the case. *(3 marks)*

f)   Explain how Computer Aided Manufacture (CAM) could be used in the construction of the case.
(2 marks)

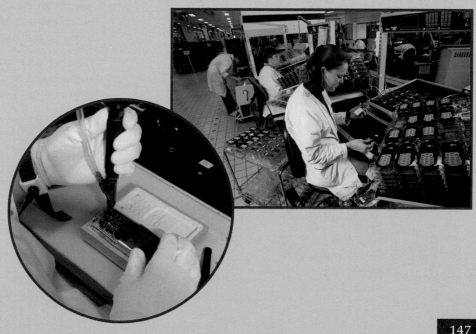

Weighing Scales

Microprocessors
(page 122)

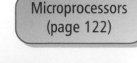

Partially sighted and blind people sometimes use specially designed equipment to help them prepare meals. You have been asked to design a set of kitchen scales that could be used by someone with poor eyesight.

Investigation

Before you start your project you will need to find out how existing kitchen scales work and what problems there are in using them.

If your school teaches food technology, ask the teacher if you can borrow a range of their scales and carry out some comparative tests. You could look at past copies of *Which* magazine to help you do this. Here are some questions to get you started:

▷ How do the scales display the weight of the ingredients in the bowl?
▷ What is the maximum weight the scales can measure?
▷ How do the scales actually work? See if you can find an old set to take apart.

You will need to find out about the problems of working with limited eyesight in the kitchen. You may be able to interview someone. Another way is to blindfold someone or give them very dark glasses and ask them to try to use a set of kitchen scales. What problems do they have?

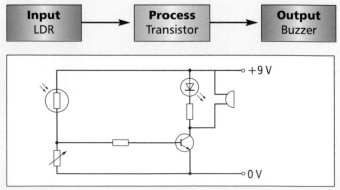

Designing the System

You can tackle the problem in two ways:

1 Your scales could be of the balance type. The weights would need to be of different shapes and could have their weight written in braille. An electronic circuit could sense when the balance was level. This could be achieved by using a reed switch and magnet.

Another way would be to use a light-sensing circuit. The scales could be designed so when they are in balance a light is sensed which switches on an audible output.

Input	Process	Output
LDR	Transistor	Buzzer

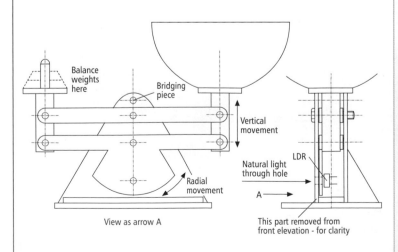

Balance weights here

Bridging piece

Vertical movement

Radial movement

LDR

Natural light through hole

A

View as arrow A

This part removed from front elevation - for clarity

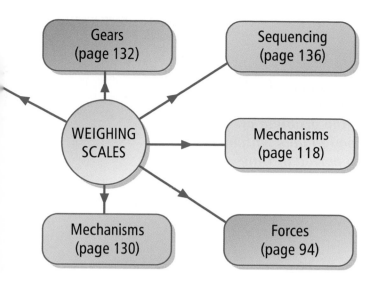

Developing the Product

Your system will need to be housed in a robust and easy to clean case. You could vacuum-form this. Will your system work all the time or will it have an on/off facility? This could be a switch or the scales could be triggered by placing the ingredients bowl on the machine.

A sensor could be switched on by the bowl. This could start the system working. A more sophisticated solution would be for this input to start the system for a set period of time, by using a monostable timer.

You could use an LDR, microswitch or touch plate to sense the bowl. How else could you start the scales working?

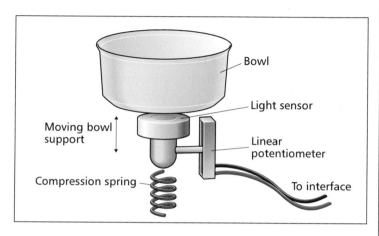

Planning and Making It!

Model your system using a construction kit. Some parts may need to be modelled using a soft material to check they work.

If you use a microprocessor then the software needs to be designed by using a flow diagram.

Check the main principles of your model work. Then produce accurate working drawings for the parts. Remember, it needs to be robust and easy to clean.

2   Another way is to use a microprocessor and interface. The interface will have an analogue input. By using a variable resistor to sense the weight of the ingredients, the computer can be calibrated to display this in large numbers on the screen. Your computer software may have a sound or voice feature, so it may actually tell the user what the weight is. The variable resistor will need to be spring-loaded, with the bowl acting against this.

Are there other ways of inputting an analogue signal into the computer?

Final Testing and Evaluation

Before you can use your product it will need to be calibrated. Use a set of weights and check your scales read the same.

▷   Use the scales yourself. Are they reliable?
▷   Are they safe?
▷   Can you clean them?
▷   Ask a food technology teacher for their opinion.
▷   Ask a person with poor eyesight to use them.

Record your findings and make any proposals for improving the product. Remember to use sketches as well as notes in your evaluation.

Automatic Pet Feeder

System Modelling
(page 64)

A national chain of pet shops has approached an electronics design company to see if it can come up with a new solution to the problem of providing a regular supply of food to pets which have to be left unattended for a period of time

Your task is to design and make a product that will supply a regular and accurate portion of feed to either a small pet or a fish over a given period of time.

Investigation

What are the most popular fish or small mammals that people need to be able to feed automatically?

What are the different requirements for feeding various breeds of fish and small mammals?

You will need to undertake some thorough research to discover:

▷ what type of feed will be dispensed – dry or liquid;
▷ how regularly it will need to be dispensed;
▷ what quantity of feed will be required;
▷ whether the feed will deteriorate.

This is likely to involve visiting local pet shops to look at existing solutions and talking to friends and relatives who own pets. A local library might have a section on pet care.

Design Specification

From the conclusions to your research you will need to write a specification which defines the requirements for the container and the electronics in terms of:

▷ its size and weight;
▷ the time spans through which it will need to operate;
▷ the safety and hygiene features needed for the pet and owner;
▷ the price range in which it would be sold;
▷ the level of reliability required.

Designing the System

Timer

How will you achieve the long time delay necessary? A typical 555 IC delay timer would not give the necessary accuracy. You could use a travel alarm clock to trigger the operation of the feeder.

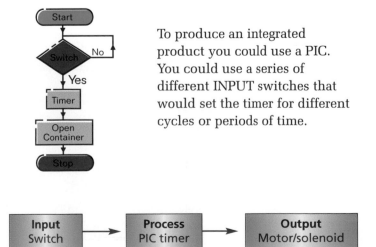

To produce an integrated product you could use a PIC. You could use a series of different INPUT switches that would set the timer for different cycles or periods of time.

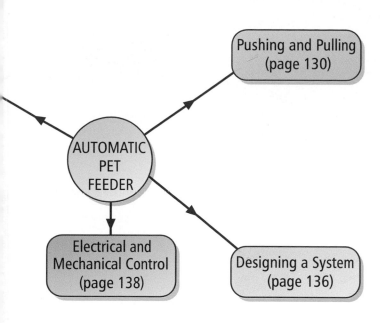

Output

What will dispense the food?
Is the feed liquid or solid?
You may need a motor or a solenoid.

Power Supply

How long will the automatic feeder run for?

Will battery power be adequate? You may want to use a low-voltage supply from a mains transformer.

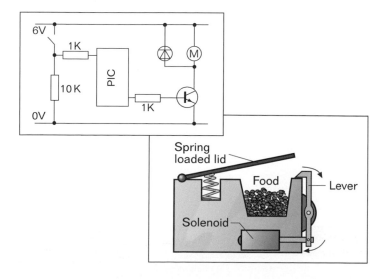

Making and Testing Prototypes

With a product of this type it may be necessary to include mechanisms to physically dispense the feed.

After sketching a number of ideas, you will need to experiment with models to check out your ideas. Technical kits are useful for testing out your ideas quickly.

Reliability could be a matter of life or death! You will need to test your design over a number of days. Data logging equipment could be useful. Record the results and make any necessary modifications.

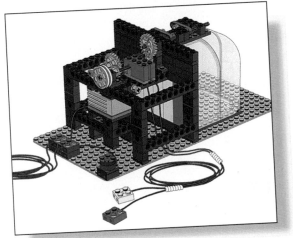

Small animal feeder

Planning and Making It!

Plan carefully the making of the electronics and casing for your final design. Make everything to the highest possible quality. How would the making process differ if you were manufacturing your design in quantity?

Final Testing and Evaluation

Before you write your final evaluation, ask friends with pets for their opinion, or perhaps take it to a local pet shop. Make sure you brief people on what they are evaluating.

▷ How well does it meet the specification?
▷ How much would they be willing to pay for such a device?
▷ How many would you have to sell at such a price to cover the costs of manufacture?

Compare your design to other similar products currently on the market.

Refer to the results of your findings when writing your final evaluation.

Index